Security, ID Systems and Locks

The Book on Electronic Access Control

Security, ID Systems and Locks

The Book on Electronic Access Control

Joel Konicek and Karen Little

Butterworth–Heinemann

An Imprint of Elsevier

Amsterdam Boston Heidelberg London New York Oxford Paris San Diego
San Francisco Singapore Sydney Tokyo

Butterworth Heinemann is an imprint of Elsevier.
Copyright © 1997 by Elsevier

 This book is printed on acid-free paper.

ISBN-13: 978-0-7506-9932-7
ISBN-10: 0-7506-9932-9

The publisher offers special discounts on bulk orders of this book.
For information, please contact:
Manager of Special Sales
Elsevier
200 Wheeler Road
Burlington, MA 01803
Tel: 781-313-4700
Fax: 781-313-4802

For information on all security publications available, contact our World Wide Web homepage at http://www.bhusa.com/security

10 9 8
Printed in the United States of America.

Contents

1 Electronic Access Control
Concepts and Study Tips

2 Ancient History
The Foundation of Access Control

3 Credentials
Cards, Codes, and Biometrics

4 Barriers
Doors, Gates, Turnstiles, and Electric Locks

5 Sensors
Information Reporting Devices

6 Computers
Software, Hardware and Intelligent Networks

7 Communications
Wired and Wireless

8 System Design
A Technical Design Perspective
by Warren Simonsen

9 System Integration
A Security Perspective
by Joseph Barry, CFE, CPP and Patrick Finnegan, qp

10 Case Histories
Noteworthy EAC Installations

Appendix

Index

Foreword

by Lawrence J. Fennelly, President
APTI, Inc. Security Consultants

I've seen the growth of electronic access control (EAC) and am happy that Joel Konicek and Karen Little have created a book to help tie the many details of this subject together.

With regard to how EAC helps protect a facility, consider this story that I heard a few years ago:

> Two security guards were sitting in the control room, casually observing their panels and CCTV monitors, when one said, "Hey, here comes Dr. Brown into the garage. The guy is like clockwork. Everyday he arrives at 7:30 AM sharp." A few minutes later the other security officer says, "This is strange. He hasn't entered the hospital yet. Why not?"

> The guards waited two minutes, then dispatched a garage-station security officer who discovered the doctor slumped over his steering wheel from a heart attack. Help was immediately called and Dr. Brown's life was saved.

My reaction to this story was — WOW! Electronic access control saved a life. Even though the guards were very alert, if they had not had the input from the wide array of communications devices that make up electronic access control, this story would not have had a happy ending.

EAC works to help keep intruders out and minimize the problems caused by key assignments. Its effectiveness in doing this is backed by statistics. As an example, I conducted a four-year crime analysis study for the two years before and after the installation of an EAC system in a college dormitory. The study showed that EAC caused a significant reduction in calls for service over the old system and an overall reduction of crime.

This book is filled with information, ideas, educational facts and case histories. I think the graphics are explicit and will aid the layman in obtaining a better understanding of the subject.

The authors, Joel Konicek and Karen Little, The Advisory Board consisting of Joseph Berry, CFE, CPP, Patrick Finnegan, Fred Coppel, qp, and Technical Editor, Warren Simonsen, have years of experience, impressive credentials and have worked hard to put together a great text.

This book is a significant contribution to our profession. It will help newcomers and even old hands better understand EAC components, computer systems, communications and networks and as well as the planning necessary to create new or improve existing security systems.

Advisory Board

Security Consultants

- Joseph Barry, CFE, CPP

  With over 32 years of experience as a Physical Security and Loss Prevention Professional, Joseph Barry is an international author and speaker on security-related subjects. He is on the Editorial Advisory Board of *Access Control and Security Systems Integration* magazine, a member of the American Defense Preparedness Association (ADPA), Executive Committee on Security, and the 1996 President of the American Society for Industrial Security (ASIS) Professional Certification Board. He is Manager, Virginia Beach Office of Scientech, Inc.

- Patrick Finnegan, qp

  Patrick Finnegan is the Manager, Security Operations for the Protection Technologies and Service Division of BDM Federal, Inc. Previously, he was Chief of the Intrusion Detection Systems Branch of the Military Intelligence BN, US Army, Fort Meade, MD. He received his B.S. degree from New York State University. He has over 25 years experience in intrusion detection and access control system design and testing. Mr. Finnegan has published numerous articles on security issues and is a member of ASIS and ADPA.

- Fred Coppel

  Fred Coppel has 26 years of project and engineering management of security and manufacturing programs for corporations and government organizations. He has been involved in all phases of security system design, engineering, installation, testing and maintenance. He has managed multiple large projects, integrating prime, subcontractors, consultants and labor force to meet client requirements in cost, schedule and quality and is a member of ASIS and ADPA.

Technical Editor

- Warren Simonsen

  Warren Simonsen is the Director of Technology for Northern Computers. He has a double BS degree in Electrical Engineering and Computer Science from the University of Wisconsin-Milwaukee and over 15 years development experience with computers and embedded control systems.

About the Authors

- Joel Konicek, President, Northern Computers, Milwaukee, WI

Joel Konicek is the founder of Northern Computers and its President since its inception in 1982. He graduated from the University of Florida-Gainseville, Fl BSEE and minor in math. He is on the Board of the Security Industry Association (SIA) and serves as the Chairman of the SIA Education Committee.

- Karen Little, President, Clear Concepts, Milwaukee, WI

Karen Little has been technical writer, illustrator, training program developer and consultant for over 20 years. She has advised numerous organizations, including the Milwaukee County Board, and founded a highly successful job training program in Milwaukee County. Today, she writes articles and produces technical documentation on a wide range of subjects, with a speciality in electronic access control.

Contributors

Guest Writers

Warren Simonsen, Chapter 8 - *System Design: a technical design perspective.*

Joseph Barry, CFE, CPP, and Patrick Finnegan, qp, Chapter 9 - *System Integration: a security perspective.*

Illustrators / Designers

Karen Little, Clear Concepts (Milwaukee, WI) lead illustrator and book designer.

Scott Darrow Illustration (Milwaukee, WI), Chapter 2 - *Ancient History.*

Deborah Dutton and Joseph Sherman Design (New Haven, CT), cover design.

Contributors

Karen Pratt (Production Manager, Butterworth-Heinemann) for final editing and publication management.

Kim Loy (Marketing Director, Northern Computers) for editing Chapters 1 through 4 during the rough draft process and ongoing help. Todd Nienhaus (Media Coordinator, Northern Computers) for coupon coordination and design.

Carl Pocratsky (Project Manager, Office of Safeguards and Security, U.S. Department of Energy), Pat French (Product Manager, Northern Computers), and Philip Little (Test Engineer, Twin Disk), for input into Chapter 5 - *Sensors.*

Leonard P. Levine, Ph.D. (Professor, E.E.C.S., University of Wisconsin-Milwaukee), Marilyn Levine, Ph.D. (founder of Association of Independent Information Professionals) and representatives from the Intel Corporation for input into Chapter 6 - *Computers.*

Ned Reiter (Communications Software Designer, Analytical Services Corp.) and Herb Guenther (Partner, Webzone Communications) for input into Chapter 7 - *Communications.*

Interviews

Tony Artrip, Director of Security, University of Miami, Jackson Memorial Medical Center. **Mitch Norton**, Director of Security, Halifax Medical Center. **Ethan Lewis**, ADT Security Systems, Inc. **Ed Loyd**, Security Director, Lam Research Corporation. **Dave Scripture**, Technology Controls. **Ken Miller**, ADT Security Systems, Inc. **Mike Ames**, Director of Security, Colstrip Project Division of Montana Power Company. **Steve Gaunt, CHPA, CFE**, Director of Security, Englewood Hospital & Medical Care Facility. **David Schatten**, Universal Security Systems, Inc.

Financial Support for Research

We would like to acknowledge that the funding and support for this book was made available by Northern Computers, Inc., a subsidiary of ATAPCO, without whose help this book would not have been possible.

Although the book received corporate sponsorship, every effort was made to keep it product-neutral in order to serve the best interests of the electronic access control industry and security concerns as a whole.

For specific product information, consult the **Security Industry Association - SIA** and the associations and publications listed in the *Appendix*.

Donations: All income generated by this book will be donated to nonprofit organizations. Contact the authors for the names of the beneficiaries.

Introduction

by Joel Konicek

Our purpose in writing this book is to promote common knowledge between *all* people who work with electronic access control (EAC).

The topic of security, by itself, fills many books and articles, and is the subject of professional seminars and classes. In the 90's, security professionals must also talk to and understand computer specialists in Management Information Services (MIS) who control corporate computer networks, security-system sales people, service technicians, communications providers and software/hardware technical support personnel.

At the same time, MIS professionals, sales people, service technicians, communications providers and technical support people **must talk to and understand security professionals**.

Frankly, it can be very difficult for the people working in unrelated specialities to understand one another.

As President of Northern Computers, a manufacturer of EAC components, I know the extent of this problem. A good portion of our training budget goes to educating and forming supportive relationships between people in various fields, in addition to developing greater technical expertise.

Here's a secret we discovered:

Sometimes a little knowledge can go a long way.

Understanding others is often based upon our knowledge of a few key concepts and our awareness of special terminology. Armed with this "little bit of knowledge," we can conversationally explore information with one another without getting lost in a technology fog.

With that in mind, Karen Little and I, along with our technical advisors, have worked hard to the identify the key concepts and terminology that we believe will help open the doors of understanding between all specialists involved in EAC.

Technology is introducing new ideas and concepts daily, but does not stamp out *"old"* ideas, at least not all at once. Consequently, we hope that this book prepares you to spot these changes so that you can integrate them into your professional life, while still supporting the everyday aspects of your chosen field.

Chapter 1

Electronic Access Control
Concepts and Study Tips

Access Control

Not too very long ago, access control was regarded as the art of keeping people out of buildings. If a stranger did gain access to a facility as, let's say, a guest of a manager, others might regard him or her suspiciously until they learn who let this person in.

Our need to know more about the people who surround us helps keep us safe. Once we become satisfied that others are OK, we relax our guard and get on with our work.

In modern society, however, transportation allows people from diverse backgrounds to gather in ways never before known in history. Our places of employment have become gigantic social mixing machines where, in some cases, several new employees are being introduced on a monthly basis, in addition to many visitors.

The result of this is that as our organizations grow larger, people lose their ability to fully know and trust one another. If a problem occurs, such as a theft or physical threat, people feel scared, intimidated and depressed.

> *Oh So True Example:*
> One corporation tripled its size within a short period of time without ever having an incident of theft. With growth, however, more and more outsiders were being brought in as employees and as guests.
>
> Suddenly, wallets and purses started to disappear, leaving the staff in a frenzy. Until a sophisticated electronic access control system was installed, work almost ground to a halt over speculation of "who did what when."

Electronic access control (EAC) is one component of a security system and it is best known for its ability to issue ID cards that replace keys. It is more than a security system, however. EAC provides an element of social engineering by quickly and securely introducing strangers into a facility in a way that they can almost instantly be trusted.

ANCIENT ACCESS CONTROL: SECURITY WAS OFTEN ADMINISTERED BY THE POINT OF A SWORD.

Chapter 1

How Does EAC Work?

This book describes EAC components, an overview of security needs and provides examples of real-life applications. Consequently, the following very brief description of how EAC works barely touches the subject, but it should provide an introduction.

EAC describes an electronic system in which information is collected and analyzed by computers. Once the information is digested, these computers issue instructions to various components, such as electronic and electromagnetic locks.

The computers have the ability to remember more information about large numbers of people than is humanly possible. The result of this is that they electronically issue commands based on the combined knowledge of:

- Security profile data
- Time and place
- Sensory data
- Management needs

MOST PEOPLE THINK OF EAC CREDENTIALS AS BEING CARDS. IN FACT, A CREDENTIAL CAN BE ANY NUMBER OF THINGS, INCLUDING BIOMETRICS, WHICH ANALYZES PHYSICAL CHARACTERISTICS SUCH AS PALM PRINTS.

In a well-designed system, security guards find these computers easy to manage. Through the use of a computer monitor, guards know who and why people are accessing the facility. They are also alerted when problems crop up and can instantly respond from their computer station by cancelling an individual's access credential and/or by locking otherwise unlocked doors.

There are times during the day when facilities need more or less access monitoring. Depending on needs, the freedom of public access might reduce the amount of monitoring during business hours, while evenings and weekends might demand more. EAC systems are programmed to adjust to these needs. They are flexible systems that take into account human behavior.

Access Points: A door monitored by an EAC system has at least one credential reader and possibly two, one for either side. It also has an electronic or electromagnetic lock and at least one sensor that tells the computer when the door is completely closed.

THESE MOTION DETECTORS CAN BE LESS THAN 3 INCHES TALL, YET THEY CAN SENSE INTRUSION OVER VERY WIDE AREAS.

This door might be surrounded with other security components, too. These include additional sensors, the most common type being motion detectors, and a CCTV system with video taping.

All these electronic devices report information (data) that help control access and provide a history of events for later investigation.

What's In This Book

This book is written in a modular fashion. You can read every page, or pick and choose topics of interest.

It provides information about:

- **Historical access control**, which provides insight into how security has changed over the thousands of years.

- **EAC credentials**, which most people think of as *ID cards.*

- **Electronic devices**, such as locks and credential readers.

- **Sensing devices**, which provide electronic feedback about what is going on at strategic points throughout the facility and its grounds.

- **Computers**, which supervise the system.

- **Control panels**, which are specialized computers that control the electronic devices, receive feedback and issue commands.

- **Communications**, which carry the electronic data between computers, control panels and the devices they manage.

- **System design concerns**, which deal with the technicalities of setting up a system.

- **Security concerns**, which deal with concepts involved in integrating all the aspects of electronic surveillance into one coherent system.

- **Case studies**, which tell real-life stories about EAC installations.

THIS IS AN EXAMPLE OF A CONTROL PANEL. IT IS A CIRCUIT BOARD AND IS OFTEN HOUSED IN AN ELECTRICAL UTILITY BOX.

THE PANEL IS A SPECIALIZED COMPUTER THAT SUPERVISES ACCESS. IT IS CAPABLE OF MAKING DECISIONS BASED ON INPUT, ISSUING COMMANDS AND REPORTING ALL TRANSACTIONS TO A COMPUTER AT A CENTRAL LOCATION, SUCH AS IN A SECURITY OFFICE.

Although issues surrounding closed circuit TV (CCTV) are an extremely important aspect of monitoring access, they are only touched upon briefly in this book. CCTV is a highly complex technology and experts, such as Charlie Pierce, have written clearly and extensively on the subject.

The information provided in this book confines itself to the direct issues of EAC and encourages its readers to combine its information with that found through other sources.

Study Tips

We hope that you will be able to quickly read and understand this book.

The amount of information you remember, however, is based on your own requirements.

If you are using this book to gain mastery of your profession — such as certification or expertise as required by a promotion — we suggest you do the following to help accelerate your understanding:

1. **Study information lists:** Lists of information, such as found in the chapters as well as the Chapter Reviews, glossaries, etc., provide a quick way to grasp new material.

2. **Dictate the information you want to remember into a tape recorder and then listen to the recording several times:** The easiest way to memorize information is to hear it spoken in your own voice. Play the tape while you're driving or perhaps when you're eating lunch. The more you make use of nontraditional study time, the faster you'll learn.

3. **Collect pictures and/or take photos:** While this book features illustrations, it does not begin to depict the wide variety of products available. Collect pictures of the latest devices and paste them into a scrapbook or folder so you always have up-to-date information on-hand.

   An even better approach is to take photos of situations that feature the concepts you're trying to learn. A significant benefit to this approach is that if you become a supervisor and/or give a talk on the subject, you'll have a wealth of visual information to share with others.

4. **Make notes in this book:** Personalize this book as much as possible. The more you highlight key passages, the faster it will be for you to review this material by skimming the highlighted areas and ignoring the rest.

5. **Take the quizzes found at the end of many of these chapters:** The quizzes at the end of most chapters are designed to stimulate your thinking and cement your memory of specific concepts.

TODAY'S ELECTRONIC AC-
CESS CONTROL USES THE
COMPUTER TO REPLACE THE
SWORD . . .

Professional Resources

Electronic access control is just one of many interrelated areas of security and facility management. Unfortunately, there are very few formal educational resources that tie together all the information you need. This means that you, like others, might have to learn through reading, other people, special seminars and sales literature.

If you haven't already done so, join professional organizations that best represent your overall needs. Make sure that they provide a good magazine (or newsletter), books, seminars, and trade shows. Sign up for their professional certification courses whenever possible.

Some professional organizations have narrow membership requirements. If you don't meet those requirements, but still need to know more about the areas they represent, consider becoming an associate member, or simply order their publications. Ask to be put on their mailing lists so that you know when upcoming seminars and trade shows are scheduled.

Self-education is a worthwhile endeavor. The money you invest in publications, seminars and trade shows will reward you with knowledge and many new contacts throughout the nation, if not the world.

Consult the *Appendix* for a list of security-related organizations and publications. Check your library or friends in the industry for additional information.

Chapter 2

Ancient History
The Foundation of Access Control

Halt! Who goes there?

"Between a rock and a hard place" is not a cozy way to describe a dwelling. Yet our ancestors in Europe, Asia, the Levant, Egypt, and North Africa often chose to live between thick walls in cramped, airless, damp, dark castles in order to protect themselves against continuous sieges.

Townsfolk feared invasions, not the witches' spells of fairy tales. Consequently, when their leaders could afford it, they built big, tough castles and city walls as a means of warding off intrusion. Then they employed dead-eyed, itchy-fingered, suspicious archers to protect the gates.

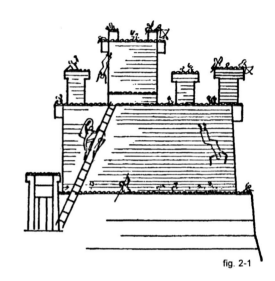

fig. 2-1

Knowing "who goes there" and deciding whether or not to let someone in was a matter of life and death. Although some castles were built on trade routes for the express purpose of imposing heavy tolls on hapless traders, their more common purpose was to protect growing civilizations and their leadership from pillage, plunder, and slavery.

ACCESS CONTROL HAS BEEN A PROBLEM SINCE THE DAWN OF CIVILIZATION.

SHOWN HERE IS THE HITTITE CITY OF DAPOUR UNDER SIEGE, AS RECORDED ON THE SOUTH WALL OF THE GREAT HYPO-STYLE HALL OF THE RAMES-SEUM. 1280 B.C.

Construction of massive walled fortresses was common. As different cultures traded and fought with one another, people were quick to note which fortification techniques were most successful. No one building style worked all the time, but many provided better than average systems of defense. Consequently, as time wore on, castles in very diverse areas acquired similar elements.

Castles were designed for controlling access, not comfort. Their success in keeping people safe for prolonged periods of time advanced society, even though eventually, most of these castles fell under siege by their enemies.

When first built, castles were usually surrounded by the sea or open land, the vastness of which provided additional protection from groups of plunder-hungry sailers and nomads.

As time advanced, small boats turned into ships and paths into roadways, providing the means to easily transport large numbers of well-armed people over great distances. These frightening travelers were often called *barbarians*, which meant "people living beyond the barriers."

The introduction of gunfire greatly reduced the defensive capabilities of castles. Consequently, castles were either abandoned, or were transformed into the luxurious, windowed palaces and chateaus that we see in travelogues. This opulent and lavish display of wealth, however, was hard won and has little or nothing to do with security.

In the early days of civilization, resources could buy fine pottery, jewelry, and cloth. They could not purchase truly comfortable abodes with windows and cross ventilation unless the peace of the surrounding countryside was absolutely secured.

This chapter provides an overview of fortified dwellings, such as castles, walled cities, and citadels. As you will learn, some of the principles of ancient access control are still used today, but not, thankfully, in such a heavy-handed way.

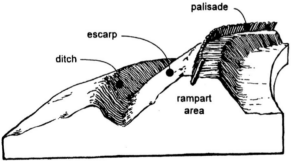

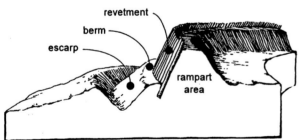

fig. 2-2

TOP: A SIMPLE EARTHWORK.

BOTTOM: A MORE SOPHISTICATED EARTHWORK. WHEN TIME AND MONEY PERMITTED, THE REVETMENT WAS MADE OF BLOCK AND STONE INSTEAD OF WOOD.

The Earth - mottes and moats

Fertile and accessible lands provide the most desirable living areas and afford the greatest opportunities for prosperity. Unfortunately, the same qualities that allow these areas to produce wealth also make them easy for others to plunder.

Growth requires the time to plan and the talent for making things come alive. Warriors, however, concentrate solely on taking things away and enriching themselves with the output of other people's labors. This difference in skills forces peaceful people to live defensively.

To mark and defend their territory, our ancestors heaved dirt and carved one or more *earthworks* into the land as a means of barricading their communities.

An earthwork consisted of at least one deep ditch that completely encircled higher territory, such as a hill or mound. The outside edge of the ditch sank at a steep angle. The inside edge (escarp), piled with the transplanted earth, created a high, sloping wall called a *rampart,* which was often topped with a *palisade* (also called a *stockade fence*).

These ditch-rampart-hill combinations transformed accessible countrysides into obstacle courses. Ditches kept invaders at a distance; often more than 30 feet. Archers braced themselves behind palisades, or on top of more sophisticated *revetments* (retaining walls), when trying to persuade invaders to move on. During peace, these ramparts served to contain livestock.

As earthworks grew in sophistication, a *berm*, which is a flat section between the ditch and the rampart, was developed. During siege, the berm caught the debris from the crumbling rampart. Without a berm, the debris could fill the ditch and create an unwanted bridge.

Julius Caesar, in the middle of the first century BC, had great experience with the use of earthworks as a means of fortification. In his diary, *The Conquest of Gaul*, he describes many fortifications and how they were defended or lost, one of which follows:

> . . . Caesar had intended to join battle; but he was surprised to see how strong their forces were and therefore encamped opposite them on the other side of a deep, but not very wide, valley. The Roman camp had a 12-foot *rampart* with a *breastwork* of proportionate height, two *trenches* 15-feet wide with perpendicular sides and three-storeyed towers at frequent intervals, joined together by floored *galleries* protected on the outside by wicker breastworks. Thus, in addition to the double trench, it could be guarded by two rows of defenders: one on the galleries, where they were less exposed on the account of the height, and could therefore hurl their missiles with greater confidence and make them carry farther; the other on the actual rampart, where, although nearer the enemy, they were protected by the galleries from falling missiles. The gateways were fitted with doors and flanked with especially high towers.

To better understand Caesar's story, refer to the glossary at the end of this chapter.

Unfortunately, ditches and ramparts were hard to maintain. Rain and wind easily eroded the slopes and battering during siege caused them to fall apart.

To reduce maintenance problems, thorny bushes were sometimes planted on the walls of the ditches so that their root systems secured the earth and their thorns posed threats to intruders. Buried within these bushes were often pointed, man-made stakes. Wood and plants, however, could be easily burned or hacked down. As more and more ditches were defiled by outsiders, townsfolk began re-engineering their ramparts by adding stone revetments to support palisades. When the materials were both affordable and available, massive stone walls replaced all wooden structures.

Despite problems, hill-and-ditch forts served their purposes and were used extensively by many cultures. They were extremely easy to create (even by an invading army), the materials were cheap (always a plus), and they gave people exercise and strong backs.

fig. 2-3

NORMAN MOTTE AND BAILEY CASTLE.

The Normans, who invaded France and England around the tenth and eleventh centuries, used earth and wood to quickly establish forts called *motte and bailey castles*. While wooden buildings were no match against battering rams and other siege engines, these devices weren't commonly used in the areas where mottes and baileys first made their appearances. Ditches, palisade-lined ramparts, and hills provided adequate defensive measures and their only weakness during a siege was a tendency to burn down.

As you can see in *fig. 2-3*, the main dwelling was built on an exceptionally high, man-made *motte* (hill) that could only be entered by means of a guarded gangplank that spanned across a deep ditch and palisade-covered rampart. Livestock, servants, and skilled labor lived in the lower *bailey* (lawn or parade ground) which was also surrounded by a rampart, palisades, and a deep ditch.

Until around the tenth century, western Europe was made up of small communities of herders and farmers. As the population grew, and more resources became available, these wooden castles were replaced with what we regard as traditional stone structures.

Walking on Water

Ditches and ramparts provided the first level of access control and because of their importance, engineers experimented with various improvements, one of which was flooding. Flooded ditches provided excellent protection when everything worked correctly. They were almost impossible to cross and kept the walls they protected from being breached by battering rams. The best were well-placed bridges concentrated the security force at entry points, thereby reducing costs and leveraging the effectiveness of a few good men.

Unfortunately, flooded ditches were not often successful except for those designed and used by the Dutch. The main reason for their failure was that ditch water froze in the winter, completely negating its strategic purpose.

In addition, stagnant stinky water often made life miserable for the castle inhabitants. It produced bacteria that nurtured disease-spreading insects. Matters were made worse when the ditch served as a latrine drain.

Last, wet ditches accelerated the deterioration of their surrounding ramparts and wooden palisades. In areas of increasing civilization and decreasing sources of wood, this posed a serious problem.

Today, we commonly associate moats with beautiful, swan-filled ponds that surround palatial estates. In reality, this type of pond was probably only maintained for decoration, as defensive wet ditches were rare. Interestingly, the term *moat*, meaning "wet ditch," was coined in England as a twist on the word *motte*, which referred to the hill above the ditch.

The Walls . . . between a rock and a hard place lies 30 feet of brick

The characteristics of castles, walled towns, citadels, and forts tended to be the same, although there were no standard floor plans. Ditches and ramparts often provided the first line of defense. The next line was a parapet that provided shielding for observation and fighting. For our purposes, the Great Wall of China provides a good overview of the parts of a defensive wall.

fig. 2-4

The Great Wall of China was one of the most effective massive barriers ever built. The wall is actually a series of long ramparts, the first of which was constructed around 476 BC. Its purpose was to mark boundaries and provide an obstacle against incursions by the Huns.

Initially, the wall was a squared-off earthwork topped by a walkway of masonry and rocks. Despite its lack of solid side structures, it was extremely effective in putting a halt to raids that hampered the Chinese silk trade, as well as keeping out the ever-invading Huns.

THE GREAT WALL OF CHINA FEATURES TOWERS, WALKWAYS, AND PARAPETS.

THE PARAPET IS A PROTECTIVE WALL MARKED BY NOTCHES CALLED "CRENELLATIONS." THE HIGH POINTS ARE "MERLONS" AND THE LOW POINTS, "EMBRASURES."

Maintenance on the wall was abandoned around 1004 and then reestablished in 1367 at the beginning of the Ming Dynasty. It was then that the wall was completely renovated to include stone and masonry sides, which helped it to remain an effective barrier until 1644, when the last Ming Emperor fell.

The Great Wall is incredibly thick. It varies from 17' and 33' wide as it winds around the mountains. In contrast, European castle walls were initially built 10' wide. Improved construction increased wall widths to 20' or more, replacing the thinner ones that easily fell to invaders.

The Great Wall stands 25' to 45' high, built atop rough, mountainous terrains. Renovations by the Ming Dynasty replaced much of its earthwork base with two solid masonry facings that had earth and gravel compacted in the center. Boulders were placed at the foot of the wall, with dressed stone rising above.

Like most castle walls, crenellated parapets rose 5' above the dressed stone of the walkway. These parapets, which are high walls on the defensive side of the walkway, are designed to protect the standing security force. Crenellation refers to the dips in the parapet that let archers shoot while still having protection. The upper part of the crenellation is called the *merlon*, while the lower part is called the *embrasure*.

The Great Wall's towers rise 12' above the parapets and are located every 200 paces. The distance between the towers allowed complete crossbow coverage throughout intervening stretches of the wall. There were no unprotected gaps where intruders could sneak through.

Its walkway is 13' wide in places and stories have it that it was wide enough to let cavalry regiments gallop along its surface, five abreast. The walkways on regular castle walls are considerably narrower, but are usually wide enough to let two guards pass each other with ease.

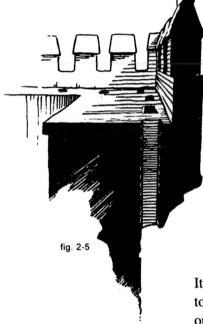

fig. 2-5

CUTAWAY SHOWING THE WALK-
WAY BEHIND CURLED-LIPPED
PARAPETS.

THIS JUTTING PARAPET CONTAINS
HOLES, CALLED MACHICOLATION
OR "DEATH-HOLES," WHICH WERE
USED TO DROP ROCKS, FIRE, AND
BOILING OIL ON THE HEADS OF
INTRUDERS.

Straight Shot Down

Throughout castledom, the construction of walls, walkways, and parapets were improved over time, although their basic shapes were similar. One improvement was the addition of a curled lip on the outside top edge of the formerly flat parapet (*fig. 2-5*). This lip was designed to deflect an onslaught of arrows back at the invaders by stopping the arrows at the edge, rather than letting them sail over the top.

Over Easy

The introduction of gunfire changed the way castles were built and ultimately, rendered castle construction useless.

Soon after cannons became one of the siege weapons of choice (around 1350), the top edges of merlons were made rounded (*fig. 2-6*). It was thought that the new shape would cause cannon balls to roll over the edges, unlike the sharp conventional edges, which would shatter and spray the guards with stone chips. In reality, cannon power simply brought the walls down. The rounded top edges provided no additional protection.

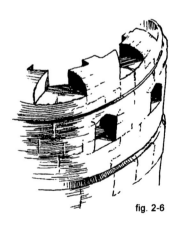

fig. 2-6

EXAMPLE OF ROUNDED MERLONS
THAT WERE THOUGHT TO DE-
FLECT CANNON BALLS BY ROLL-
ING THEM OVER THE SURFACE.

Beneath the Walls

Combative mining (tunneling) was a very effective way for the enemy or the defenders to do damage to one another's property. Miners,·for example, could secretly spirit away dirt from under towers, causing those towers to collapse, creating a breach (hole) in the wall. (See *fig. 2-14, item 2.*)

Most castles in the Levant used rounded towers to offset mining's effectiveness. Although square towers provided more spacious living quarters, they were less stable when the earth was removed from under their corners.

Europeans initially preferred square towers until mining became popular in their region. Consequently, as European castles were enlarged or remodeled, they added rounded towers to their curtain walls, creating mixtures as is seen in the plan shown in *fig. 2-10.*

More Than Garden Material

Masonry and/or stone, of course, made the best walls, but earth still played an important part, especially when the lack of wealth forced the community to cut corners, or stone wasn't available. Even with sufficient resources, earth was still mounded behind the walls to provide greater support during times of siege.

Another good defense was to have earth compacted in front of a masonry wall, because earth could better buffer the blows of increasingly popular gunfire. Unfortunately, this type of front was difficult to maintain and if the siege continued long enough, would fall into the ditch, providing a bridge where none was wanted.

Boiling Oil

Perhaps you've heard about the anointment of boiling oil on intruders' heads. Well, when castle walls were first developed, parapets allowed archers to safely shoot forward, but did not provide protection for them to shoot downward. This was particularly troublesome when the enemy attempted to climb up the wall.

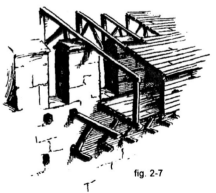

fig. 2-7

To fix this problem, holes were carved in the outer walls near the base of the parapets (*fig. 2-7*). Heavy timber beams were stuck into these holes and protruded outward like fingers. At time of siege, timber was placed across these beams, making a temporary walkway, wall, and roof, creating a *hoard*. When the guards wanted to release rocks, fire, or hot oil on invaders' heads,

HOARDS PROVIDED A TEMPORARY COVERED WALKWAY THAT PROTRUDED AWAY FROM THE CASTLE WALL, ALLOWING THE GUARDS TO FIRE STRAIGHT DOWN WHILE STILL BEING PROTECTED.

all they had to do was man the hoard, lift a beam, then bombard the payload down through the open space.

Today, you can often see these holes running along the tops of the walls and towers of ancient castles. Wooden beams, of course, have long since rotted. A structural refinement of hoards, called *machicolation* (*fig. 2-5*), made stone-work outcroppings part of the wall itself. The activity on a machicolated walk-way inspired a more popular name for these types of openings — *death-holes*.

The Star of Defense

The last great forts and walled cities pioneered by the Spanish and Dutch, and later adopted by France's military engineer, Sabastian le Prestre de Vauban in the 17th century, substituted *bastions* for rounded towers.

A bastion is a five-sided tower. Its back side is part of the curtain wall. Its front two sides meet at an angle and protrude from the tower like a point. This un-

FRENCH STAR-SHAPED FORTS WERE PROTECTED BY BASTIONS AND FREE-STANDING BUILDINGS CALLED BREASTWORKS AND RAVELINS.

THE BAILEY IN THE CENTER OF THE FORT WAS WHERE PEOPLE LIVED AND WORKED.

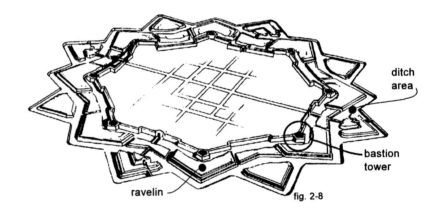

ditch area

bastion tower

ravelin

fig. 2-8

usual shape allowed gunners stationed in the wall between the towers to shoot closely along the flank of the wall without fear of hitting the tower itself with friendly fire.

Vauban improved his forts by adding low, triangular buildings, called breast-works and ravelins, in the ditches that surrounded the bastion towers and main wall. These freestanding buildings served as observation decks, firing points, and obstacles, which helped maintain France's strength until the revolution.

Keeps and Donjons

Castle living quarters were known as *keeps* or *donjons*. In all probability, the word "dungeon," which is known as a dark, gloomy prison, evolved from the term "donjon." Living quarters in the good old days were not bright or airy.

In the Levant, stone and block castles were common. Their thick walls and building style had been thoroughly tested during thousands of years of warfare and slavery. Castle construction in other lands, however, was based on what the terrain offered and the availability of skilled labor. Although Scottish stone keeps called *duns* and *brochs* (*fig. 2-9*) existed, the invading Normans were too busy fighting in France and England to build anything more substantial than wooden motte and bailey castles (*fig. 2-3*).

fig. 2-9

SCOTTISH BROCH

In the eleventh century, after the Normans settled their affairs, the need for stone castles created a building boom. These castles included massive rectangular keeps, built on flat ground near the curtain wall inside the bailey. The White Tower of London is an example of this style.

Stone keeps were more commonly built as an integral part of the curtain wall, because their weight demanded a very firm and deep foundation (*fig. 2-10*). It was more efficient to create a strong foundation in conjunction with a heavy curtain wall, than to build one separately in the center of the bailey. In addition, the depth of the keep's foundation helped slow down invasion through mining, although it didn't stop it.

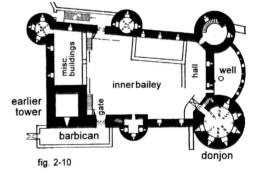

fig. 2-10

The ancient keeps in the Levant, however, had rounded exterior walls. Rounded walls were more resilient to battering, stone barrages, and mining operations. During the eleventh and twelfth centuries, western Europeans, called Crusaders, successfully invaded the Levant and built castles with Norman-style rectangular keeps. Unfortunately, siege weapons in the region could quickly reduce rectangular buildings to rubble, so the Crusaders adapted the cramped-but-secure rounded-style for their own protection. When the Crusaders returned to their homeland, they introduced rounded keeps, and from the twelfth century on, the west adopted this style.

PLAN OF THE CHATEAU DE NAJAC. THIS SHOWS THE OLD SQUARE TOWER AND THE ADDED ROUND TOWERS CALLED "DONJONS."

THE KEEPS WERE THE LIVING SPACES BUILT WITHIN THE THICK WALLS.

THE RECTANGULAR CUTOUTS IN THE THICK WALLS REPRESENT ARCHER GALLERIES.

Keeps protected people and foodstuffs and were, therefore, highly guarded. The main entrance opened one or two floors above the ground, accessible through heavily guarded, removable staircases protected by gatehouses called *barbicans*. Access to the ground level was through an interior, often circular, stairway.

Inside the limited space of a keep was room for the lord or commander and round-the-clock guards. A chapel was often included to keep up morale. Lower floors held the stores and weapons. Once gun powder was perfected, of course, ammunition was stored in separate quarters.

Castle roofs were flimsy compared to their walls. Consequently, the outside walls of the keep towered high above the gutters in hopes that the enemy's stone- and fire-throwers wouldn't have enough reach. Some roofs were cone-shaped and coated with metal so that fire would slide off. Unfortunately, as weapons became more powerful, these high walls no longer offered protection.

In time of battle, a secret exit called a *sally port* was used. Unfortunately, early keeps only had two exits — the front door and the secret passage — making it easy to trap the residents. As time wore on, multiple sally ports were created to hasten escape.

The Windows . . .
forget the drapes and pass the ammunition

Until the fifteenth century, windows were extremely small. The windows that did exist were heavily shuttered with strong timber and metal bolts. Those who used larger windows tended to be murdered in their beds.

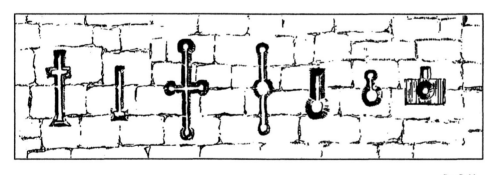

fig. 2-11

The placement of openings in exterior walls was always seriously considered. As part of siege preparations, the enemy would thoroughly examine the exterior walls before attacking, looking for weaknesses. In 1203, surprised residents of the Chateau Gaillard in Normandy lost their castle due to an ill-placed opening. The determined enemy gained access by crawling up a latrine shaft!

WEAPON-SUPPORTING OPENINGS IN CASTLE WALLS WERE MODIFIED OVER TIME.

THIS ILLUSTRATION SHOWS BOW AND CROSSBOW SLITS AS WELL AS SLITS MODIFIED FOR RIFLE HOLES.

Openings in castle walls were created primarily for defense. Until the early twelfth century, those openings looked like slits. Behind the slits were sloped galleries big enough to fit an uncomfortable, irritated, straight-shooting archer looking for something to do.

When crossbows were introduced in the twelfth century, existing slits were modified to include a horizontal cross-section. Although crossbows were generally too heavy to use in mounted battle, they were deadly when braced against a wall. The slits were modified again in the fifteenth-century to accommodate rifles.

In their time, however, crossbows were so effective that by 1139, they were prohibited, because the wounds they inflicted were considered to be so barbarous. Given that horrible assessment, crossbows were immediately enlisted in all defensive warfare until guns became popular.

The Locks . . .
gates, trap doors, and death-holes

Today we ceremoniously give the "Keys to the City" to visiting dignitaries. In ancient times, bestowing these keys meant that the defending garrison was marching out in defeat, lead by their commander, who just turned the "city keys" over to his enemy. This act was called *capitulation*.

fig. 2-12

BRIDGE PROTECTED BY THREE GATEHOUSES.

Castles were situated to provide maximum protection. Thick stone walls were nestled against mountains and riverbeds whenever possible. Walled cities were planned for the maximum movement of their security force. An open marketplace was used as the rallying point for the troops during siege and was generally placed in the center of town. Roads leading from the marketplace went straight to gatehouses and walls.

To limit access, bridges provided the best protection. It was not uncommon to see stone gatehouses at both ends, with several smaller ones standing in the middle (*fig. 2-12*).

Gatehouses were miniaturized castles. Small galleries inside the walls provided the space for archers (*fig. 2-10*), plus arrow or crossbow slits. They often had machicolated parapets and enough death-holes in their ceilings to annihilate trespassers (*fig. 2-5*).

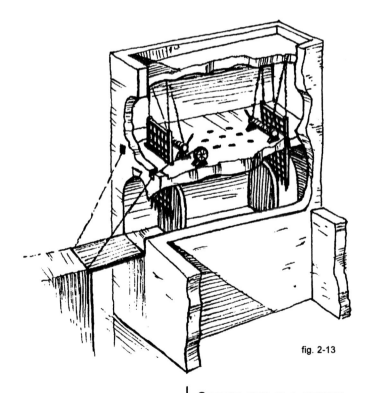

fig. 2-13

CUTAWAY VIEW OF A GATEWAY, INCLUDING A DRAWBRIDGE, SETS OF PORTCULLISES AND HEAVY DOORS, AND DEATH-HOLES.

Drawbridges were often used to traverse dry ditches and moats. They could be quickly raised and had the added advantage of dropping enemies into unwanted areas, such as deep pits. If enemies managed to enter a gatehouse, they would find themselves trapped between grated metal *portcullises*, which could be quickly dropped or raised, as needed (*fig. 2-13*). Several sets of portcullises were often used throughout castle passageways.

The doors themselves were made out of strong timber, laced with metal grillwork. Unfortunately, while wood was the most resilient material available against battering, it could be easily torched.

Death-holes in the ceilings were used as often for dousing fires as they were for firing on invaders. Obviously, a raging fire could instantly open all the doors, but even a smoldering fire could produce enough acrid smoke to cause the guards to abandon their positions.

Once the enemy passed through a gate, he still faced a maze of interior walls, laced with secret passages. Internal stairways were extremely narrow, made to limit the rush of invaders. Even the ascending angle and turns of the staircases were designed to hamper the combat-power of enemy troops by inhibiting their right arms. Security guards, stationed in galleries at every level, fervently hoped that, in case of a siege, the bodies of the enemy crammed in this tight space would provide a dam against further infiltration.

The Keys . . .
shovels, ladders, and battering rams

Castles were built to keep people safe. Although leaders often planned their own wars while sitting behind the walls, the castles themselves were designed to allow people to go about their daily tasks, free of fear. Their strength, however, made them targets of attack until the ferocity of guns and air power ultimately reduced their usefulness.

Sieging armies had four main plans of attack. They could:

1. Surround the castle and hope that starvation, thirst, disease, or boredom would provoke a surrender.

2. Climb over the wall (*escalade*).

3. Batter the wall to create a break (*breach*).

4. Dig ditches up to the wall (*trenching*) in which the troops could sneak forward, or dig a tunnel under the wall, causing the wall to collapse (*mining*).

Armies in the Levant used all the above methods thousands of years before they were introduced in Western Europe. Escalade was the most terrifying. It was known in Old Testament times and its last recorded use was in 1945 at the taking of Fort Drum in Manila Bay.

Escalade, as seen in *fig. 2-1* on the first page of this chapter, required the enemy to place a ladder on the side of the wall, climb up, and engage in hand-to-hand combat. To counter this, the guards showered the invaders with stones, arrows, and, of course, boiling oil.

fig. 2-14

A siege tower, also know as a *penthouse*, was a mobile tower built as high as the target wall. The enemy would pack their best troops into the tower and then pushed it close to the wall for better firing advantage as well as escalade. The defending guards, however, could tunnel under their own wall, excavate a timber-lined pit beneath the siege tower, and then set the timbers on fire, thereby weakening the structure under the earth and causing the tower to collapse under its own weight (*fig. 2-14, item 3*).

Battery was used to gain the advantage by breaching a wall. Enemies would attack a wall by battering it with a ram. Of course, by working under the guards' noses, they tended to quickly perish while doing their job. To protect and house the ramming operations, smaller versions of penthouses, constructed of wood and covered with metal (when available) and wet hides to ward off fire, were pushed against the walls.

Battery grew more efficient through the use of the *trebuchet* (*fig. 2-14, item 1*), which featured a sling that could throw stones at the side of the wall. More commonly, it was used to fling fire, dead animals, or live prisoners over the top. Ultimately, cannons were used and, as they were refined, could bring down

#1 - A TREBUCHET BEING FIRED FROM THE INSIDE OF A FORT.

#2 - A TUNNEL BEING DUG FROM THE INSIDE OF THE FORT LEADING TO AN AREA UNDERNEATH THE SIEGE TOWER.

#3 - A TEMPORARY ROOM (GALLERY) SUPPORTED BY TIMBER AND FILLED WITH FLAMMABLE MATERIAL. WHEN BURNED, THE GROUND UNDER THE SIEGE TOWER WOULD GIVE WAY.

#4 - A MULTI-STORY SIEGE TOWER.

a wall in hours, something that once took other battering devices weeks or months.

Incredibly, one of the best weapons was the shovel. Enemies dug deep below the earth to undermine castle walls, causing towers to crack and sink. The easiest structures to breach were rectangular in shape, because the entire structure would tip once a single corner was undermined. Consequently, most castles were designed with more stable rounded towers, which, as fortune would have it, were cheaper to build because they didn't require fine joining work.

The guards could detect secret mining operations by placing a dried pea or pebble on the skin of drums strategically located around the inner wall. Dancing peas alerted them to mining activities, at which time, they'd man their own shovels and intervene by counter-mining. The resulting fighting in the dark bowels of the earth was both fierce and frightening.

Mining activities usually favored the guards. Not only did they sabotage enemy tunnels, they also dug traps under the land approaching the walls. These traps consisted of hollowed-out, earth-covered areas that were strong enough to support foot traffic, yet weak enough to collapse under the weight of a traveling siege tower.

Last, trenching let the enemy get closer to the castle walls by cover of earth and is used in warfare to this day. The trenches provided a protected path along which the troops would position themselves to rush the wall once a breach was made by the rams or cannons. Trenches also provided the protection grenadiers needed to lob hand grenades. Prior to the refinement of gunfire, these hand grenades often did as much damage as cannons and could cause far more terror.

The Best Defense

The best defense, one that was used by all great civilizations of the past, was to provide a civic structure within which citizens and neighboring communities could get along.

Trade, good communications, and just laws have shaped the world into what we know today. This defense has always included the intelligent use of information, the ability to be inventive, and the flexibility to change.

Modern access control serves to protect homes and businesses by providing more information about "who goes there," and welcomes most comers. The key word, of course, is *access*. It is access that gives us the power and means to grow with the times.

fig. 2-15

Chapter Review

Many of the concepts used in castle construction are still used today. The following glossary of terms might prove useful, especially if you are interested in fortification.

Bailey: the courtyard inside a fort. Also refers to the yard inside protective barriers.

Barbican: a gatehouse or enclosed entrance that protects the gate to a walled city, an exterior door or an approach to a bridge. It houses guards and provides observation areas, places to fire weapons, portcullises and possibly death-holes through which objects can be dropped on the heads of intruders.

Bastion: a five-sided guard tower, where the back side is a curtain wall (main wall of the fort) and the front forms a protruding point. The forward angle of the bastion's front denotes the salient angle of weapon fire. This angle provides the most coverage along the flank (side) of a curtain wall from a point between towers without risking damage to the bastion itself from "friendly fire." (See *fig. 2-8.*)

Battery: an area that houses big, battering weapons, such as rams, cannons and mortars.

Berm: a cleared space between the top of a ditch and the foot of a rampart. It is designed to contain the debris from a battered rampart and keep them from falling into the ditch, making an unwanted bridge into the fort. (See *fig. 2-2.*)

Breach: an opening made in a rampart or wall by battering, mining, fire and/or explosives.

Breastwork: a wall that sticks out of the curtain wall that protects the guards against enfilade fire. (See *fig. 2-8.*)

Capitulation: acknowledgment by the defenders of a fort that they lost the battle, followed by agreement to turn the "keys to the city" over to the enemy.

Circumvallation: a trench, cut by the attackers around a fort they are attacking, which isolates that fort and deprives it from relief.

Citadel: a fort or castle that defends a walled town. It usually sits on the most commanding ground inside the curtain wall and, if the town is defeated, provides a retreat for the guards.

Command: the ability to dominate an area by observation or firepower by virtue of being situated on the highest elevation.

Covered Way: a walkway around the outer edge of a ditch, protected from behind by a manned parapet and in front by a glacis (cleared park or field). It allows troops to be protected from behind while being posted at the first line of defense.

Crenelle: a notch cut into a parapet (wall) to form a battlement (battle station), which gives protection to the guards engaged in observation and/or weapon fire. (See *fig. 2-4.*)

Curtain: the main defensible wall of a fort or castle.

Ditch: the excavation surrounding a defensive area, usually backed by a rampart on the inside edge. The ditch may be dry or flooded. Also called a "moat."

Donjon: a French term for "keep," which is where people live in a castle.

Dungeon: a type of a jail found in a castle, which is sometimes located in a pit beneath a tower.

Embrasure: the lower part of a notch in a crenellated parapet (wall). (See *fig. 2-4.*)

Enfilade: weapon fire directed in such a way that the projectiles travel along the length of a wall or line of troops.

Escalade: the act of climbing a wall by means of ladders or siege towers. (See *figs. 2-1* and *2-14.*)

Escarp: the inner wall of a ditch below the rampart. (See *fig. 2-2.*)

Flank: the side of a thing. Can refer to the side of a curtain wall or a line of troops.

Gallery: a room. When located inside a curtain wall, it usually provides an observation area and a place to secure a weapon. (See *fig. 2-10.*)

Glacis: a stretch of cleared land or park on the open country side of a ditch, which keeps strangers in plain view as they approach the castle.

Insult: the taking of a castle by surprise, without formal siege.

Keep: the stronghold or residential part of a castle.

Machicolation: holes in the ceilings or protrusions from the exterior walls from which missiles or other matter are dropped upon the intruders below. Also called "death-holes." (See *fig. 2-5.*)

Merlon: The upright section of a notched parapet (wall). (See *figs. 2-4* and *2-5.*)

Moat: a deep ditch, either wet or dry, used as an obstacle to define and defend an area.

Motte: a mound (hill) surrounded by moats, palisades, keeps and a bailey that is topped off by a defensible dwelling. Mottes formed the landscape for early Norman castles. (See *fig. 2-3.*)

Palisade: a pointed wooden fence made out of sharpened stakes. (See *fig. 2-2.*)

Parallel: in military terms, refers to a ditch that runs parallel to the wall of the fort, which provides protection for advancing troop movement.

Parapet: a wall on top of a rampart or building behind which guards observe the surrounding countryside. Also known as a *breastwork.* (See *fig. 2-4* and *2-5.*)

Portcullis: an iron grill door that can be lowered like an *overhead garage door.* Strategically placed in gatehouses or towers in front of heavy wooden doors, they are lowered as necessary to form additional barriers and traps. Iron grills are used to withstand the fire and chopping that might be directed against wooden doors. (See *fig. 2-13.*)

Rampart: a bank of earth that stands on the fort-side of a ditch. It is usually created from the earth excavated out of the ditch and serves to provide the necessary command (height) to the parapet. (See *fig. 2-2.*)

Ravelin: a low, freestanding, triangular building that is placed at intervals along the ditch that runs around a curtain wall. It enables defending gunners to bring cross-fire to bear on the ground around the fortress. Popular with the Dutch and French, fort-plans that use ravelins looked like multi-pointed stars. (See *fig. 2-8.*)

Retrenchment: a line of defense within a fort, made up of additional ditches and ramparts, that provide fall-back positions in case the main wall is breached.

Salient: the path of a projectile that provides the greatest defense along the side of a fort; i.e., the projectile's travel hugs the wall without hitting the curtain wall and its protruding towers. The *salient angle* describes the crossing lines of fire that cover the greatest range along the curtain wall with the minimum of dead (non-defendable) area.

Sally Port: a small, usually hidden, tunnel, gate or door leading out of a fort which allows people to escape during time of siege or *sally forth* (sneak out) to attack the enemy.

Sap: temporary trenches dug during sieges used to keep the advancing troops (whether defenders or enemies) protected as they move about.

Siege Tower: a temporary, multistory structure built by the attackers, which is placed as near as possible to a curtain wall. It provides a protected place to fire weapons as well as the necessary height to mount the wall and invade the castle. (See *fig. 2-14, item 4.*)

Trous-De-Loup: a series of pit traps, each about 6.5 feet deep and wide, used to form an obstacle course around the glacis (exterior park). They hold sharpened stakes in their bases and were often covered by earth-covered roofs that are strong enough to support pedestrian traffic, but not horses, wagons or siege towers.

Chapter 3

Credentials

Cards, Codes, and Biometrics

The term *"credentials"* refers to documents that verify a person's identity. People who present their credentials to officials or check points are regarded as being *authentic* . . . they are who they say they are.

In electronic access control (EAC), credentials refer to cards, tokens and physical patterns, such as fingerprints, that identify people. Cards and tokens are presented to media readers for *authentication*, while physical patterns, such as fingerprints, are *verified*. When a credential is *validated*, access is granted.

A credential is regarded as secure if it strongly resists alteration and/or duplication through forgery, or illegal use gained through spying. A secure credential when used by itself, however, does not resist use by an unauthorized person.

To increase the likelihood of truthful authentication, a single access transaction requires a multiple-step verification process. This process often combines a card with other identifiers, such as a personal identification number (PIN), biometric feature (such as fingerprints) and even photo/video identification.

Not all access events require the same level of security. Exterior doors, for example, usually require more validation transactions than do interior doors.

> *Example:* In a highly secure chemical plant, check-in is monitored by guards. Employees display and use their proximity photo ID cards, punch in their PINs, and have their palm prints read and verified. All these transactions take around 27 seconds per person, including a bit of gossip.
>
> Once inside the plant, however, employees wearing their proximity cards move unhampered from monitored area to monitored area. This is because their proximity cards provide hands-free validation. The purpose of the internal EAC system, then, is to track "who goes there" and when they do it. It is not intrusive.

The most common cards or tokens used throughout the world are based on Wiegand, magnetic stripe and proximity technologies in conjunction with PINs. These technologies, plus more, are described later in this chapter.

CARDS OFTEN REQUIRE PERSONAL IDENTIFICATION NUMBERS IN ORDER TO VERIFY THE AUTHENTICITY OF THE USER.

CARDS THAT ARE MORE SECURE CAN BE USED IN MULTIPLE APPLICATIONS.

THE READER/KEYPAD UNIT ILLUSTRATED IS PART OF A SYSTEM DESIGNED TO COLLECT TIME AND ATTENDANCE INFORMATION BY USING A STANDARD EAC CARD.

Management Procedures

No matter what credential technology you use, a facility is only as secure as its credential users' honesty. To manage a well-run system, then, you need to establish procedures that will verify identification, credential usage and termination, as the following overview describes:

Enrollment Procedures:

These let you enter data on access entitlements for users of the system, time zones, access levels and geographical controls (identifying buildings and doors). Periodically, you'll need to update your information, including addresses, promotions, etc.

Tracking Procedures:

These check to make sure that the credentials of your users are still in the system and are not altered or worn in any way. Plan on reissuing credentials and PINs at staged intervals. In a large system, for example, it would be disastrous to discover that all magnetic stripe cards wore out around the same time.

Termination Procedures:

These make sure that EAC authorization can be stopped the *moment* the user is terminated. In addition, they make provisions to retrieve outstanding credentials even though those credentials are no longer active in the system. This reduces chances for counterfeiting.

According to the *Handbook for Employers*, issued by the U.S. Department of Justice, the following documents help establish identity:

United States Passport

Certificate of U.S. Citizenship

Certificate of Naturalization

Unexpired Foreign Passport

Alien Registration Receipt Card, including photo

Unexpired Temporary Residence Card

Unexpired Employment Authorization Card

Unexpired Reentry Permit

Unexpired Refugee Travel Document

Unexpired Employment Authorization Document

Drivers License (U.S or Canada)

ID Card issued by federal, state or local government

School identification card with photo

Voter's registration card

U.S. Military Card, or draft record

U.S. Coast Guard Merchant Mariner Card

Native American Tribal Document

Enrolling Credential Users

A dishonest person with the best credential will cause harm to the facility. An honest person, free to roam, is a blessing.

> Screening people prior to issuing credentials is important.

Validation: The very lowest level of screening is a visitor's pass issued on the say-so of another person in your organization. These passes do not unlock doors, but they do provide a way of identifying a stranger who is walking through the halls.

The more cautious you wish to be, the more *verified* information you need to collect and maintain. The following list sets forth some considerations:

1. Name. This low-level verification is based on a recommendation or casual identification, such as a business card.

2. Name and address as *verified* by a driver's license or other officially recognized credential. (See previous page for list.)

3. Background information (automobile license plate, health records, family names, etc.).

4. Employment information (employer, employer address, department, title).

5. Military information (if any).

6. Arrest information (if any).

7. References (employer, neighbors, family, civic, etc.).

The information you gather is only as accurate as your ability to ***double check its authenticity***. Failure to check information, even when a well-ordered portfolio is easily at hand, can have disastrous results.

Enrollment Time: Depending on your needs, the time it takes to enroll new people into your system is a factor in deciding what types of credentials you need to issue. It takes longer, for example, to customize a magnetic stripe card with a color photo than it does to issue a card from a stock pile.

Your enrollment procedures must calculate this processing time — minutes or days — so that you can keep up with your enrollment volume.

> You can't keep people out just because you have no time to let them in . . .

Encoding Considerations: Although most people think of credentials in terms of names and identity, computerized EAC systems associate people with code numbers. When used in cards or tokens, these codes are hidden from people (they are not PINs).

With regard to codes, you need to:

- Know the total number of codes available in your system and how that total affects your future needs.

- Decide whether you want to customize those codes or accept standard numbers from the credential manufacturer.

- Know whether you can link PIN codes to your card or token codes.

Of the most popular cards, magnetic stripe, proximity and bar-coded cards can be customized by the user, but Wiegand cards cannot.

When you depend on manufacturer-encoded cards, you must keep a sufficient number of those cards on-hand to meet your enrollment demands. If you do not customize these cards with the user's name or photo, these cards can be reassigned until worn out.

For a number of reasons, most systems impose restrictions on the total number of codes available. Keypad systems, if not chosen with care, can be quite restrictive.

Coding is dependent on the electronic circuitry of the media-reading mechanisms, memory, software and, in the case of keypads, available keys and internal switches.

To increase the number of codes available, some cards provide a *facility code*. This code is placed before every single number in a standard range of codes, thereby increasing the total number of codes available. Example:

> Codes 1 through 9,999 are available to approximately 10,000 users. By using three facility codes with this range (say, 1001, 1002 and 1003), you increase the total to approximately 30,000 available codes. (See details under "Magnetic Stripe Cards" later in this chapter.)

Control panels validate the facility code first, then the credential code.

When designing a system, you want your control panels to interpret the facility and card code whether or not that panel is linked to a main computer. When a control panel is dependent on a main computer and the communications link between them is broken, that panel might accept all valid facility codes *without* checking card codes. Worse, the panel might not function at all.

Credential Readers

Credential readers (as well as scanners, keypads, etc.) act as the "middle men" between the credential user and the control panel. They are stationed at one or both sides of a protected door.

When two readers are used to protect both sides of a door, the system enforces *antipassback* procedures, which discourage people from sneaking into areas. If a credential user fails to use the readers in the right sequence on both sides of the door, an alarm is sounded and guards are notified with a message displayed on their control monitor announcing who made the mistake.

All readers send credential codes to a control panel. The control panel compares the code it just received to existing databases. Depending on what the panel finds, it issues instructions to open a lock, sounds an alarm, or does nothing. If the control panel is dependent on a main computer, the computer checks the code, then sends validation information back to the panel and the panel takes it from there.

Readers can take many forms. The most common are **card swipe** or **insertion devices**, **proximity devices** (based on radio frequencies), **keypads** (usually 4-key or 10-key), **scanning devices** (also called "optical readers"), and **sensing devices** (used in biometrics).

With the exception of biometric scanners, most readers are used for a variety of commercial applications in addition to EAC. Bar code readers, for example, automate pricing and stocking information in department and grocery stores. Magnetic swipe readers are used in charge and debit card transactions.

One increasingly popular use of readers is to *track time and attendance* for payroll. The same credential that lets an employee access a facility also aids in calculating his or her paycheck — and does so far more accurately than payroll systems compiled by time clocks and cards! The result can be very cost effective.

EAC readers differ from commercial readers because they require *tamper-resistant mountings*. Access to wiring by removing a poorly mounted reader can render an electronic lock useless. Tampering and vandalism, not card duplication or fraud, account for a significant percentage of reader failures!

Proximity readers, which can be completely hidden within a wall, are the most tamper-resistant. Others, depending upon decorating needs, can be surface mounted or embedded, but must not have exposed screws or prying areas. Fortunately, reader tampering can be detected by attaching sensors to the reader mount. Once an EAC system detects tampering, it can ring bells and signal authorities.

Chapter 3

Cards and Photo IDs

The increased speed and storage capacity of computers, coupled with decreasing prices for computer equipment, printers, video cameras and digital cameras is making the creation of computerized image databases and photo ID cards easy and inexpensive.

While traditional photos and Polaroid technology are still being used to apply photos to cards, digital imaging provides:

1. Truly instant image creation (no chemicals or waiting period).

2. Instant computer files (no scanning necessary when the camera is connected to the computer).

3. Tamperproof, easily produced ID cards in color or black and white that don't require lamination.

4. Complete control over the design and processing of ID cards.

Full color photo ID cards provide better security because they contain more visual information than black and white images. In addition, a full color image database on a computer provides excellent verification resources for the authorities who need to make visual comparisons.

At this writing, photo ID cards are most commonly produced by the following methods:

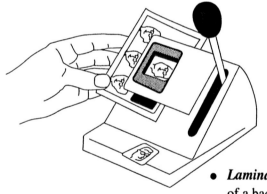

THIS ILLUSTRATION SHOWS A MACHINE DESIGNED FOR CUTTING PHOTOS FROM A SHEET. ONCE CUT, THE PHOTO IS GLUED TO A CARD, THEN THE WHOLE CARD IS LAMINATED.

PRINTING DIRECTLY ON A CARD IS RAPIDLY REPLACING HAND CUTTING AND LAMINATING.

THE SURFACES OF PRINTED CARDS ARE THINNER WITHOUT THE LAMINATION AND ARE LESS LIKELY TO JAM MAGNETIC STRIPE READERS.

- *Lamination:* In this process, a photo (traditional or Polaroid) is cut out of a background and pasted on a card, which is then covered by sheets of clear plastic.

- *Dye Sublimation:* In this process, a full color digital image is printed on a card through a process called *dye sublimation,* which works as follows:

    The image is reproduced by placing a tightly spaced series of dots on a vinyl card. If color is used, these dots are made through cyan, magenta, yellow, and black ribbons. Print processing uses variable heat temperatures to melt the colored dots, blending their hues and producing a wide-range of colors. When these dots cool, they permanently bond to the card surface, making the card tamper resistant.

- *Black and White Laser:* In this process, special properties in the cards' material bonds with black laser printer toner.

Card Size

What you put on a card, of course, is limited by the size of the card itself. To increase cost-effectiveness and to promote multiple technology cards, the EAC industry is striving to standardize all cards in terms of size and thickness. Two standards, which were developed by the banking industry, are:

- CR-80: most common credit-card sizes
  (2.125" tall by 3.375" wide by .03" thick)

- CR-60: slightly taller than the CR-80
  (2.375" tall by 3.25" wide by .03" thick)

CR-80 is the most common size and fits all card swipe and insertion readers. CR-60, however, fits all card swipe readers, but not insertion readers. As swipe and insertion readers perform exactly the same tasks, it is important to be aware of the CR-60's reader limitations when designing a system.

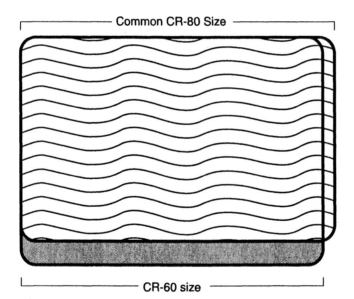

Credential Types

The following pages provide a background on 12 basic credential types which are listed alphabetically below.

Some credentials can be easily customized — important for facilities that want a lot of control over information — others depend on supplier-issued ID codes.

The credentials you select depend on your need for easy customization and your overall security goals. Supplier encoding, while often being regarded as highly secure, is not necessarily the best. Combining technologies (such as photo identification, magnetic stripe and a PIN) can result in very satisfactory credential security.

Credential Types	Related Technology
Bar coded cards and objects	Light and dark patterns interpreted by optics
Barium ferrite cards	Magnetic pattern
Biometrics	Physiological patterns interpreted by various means
Hollerith cards	Holes that allow the passage of light or electrical current
Infrared cards	Light and dark patterns interpreted by optics
Keypads	Keyboard input
Magnetic stripe cards	Magnetic media
Mixed-technology	Combined technologies
Optical cards	Light and dark patterns interpreted by optics
Proximity cards and objects	Radio transmission and computer chips
Smart cards	Computer chips
Wiegand cards	Magnetic pattern

Bar Coded Cards and Objects

Bar codes are seen as a set of parallel thick and thin black lines. These lines form a light/dark pattern that is interpreted by an optical reader or scanner as a code number.

Currently, there are more than 13 different bar code symbol sets, plus variations on these. The most popular codes include the Uniform Pricing Codes (UPC-A, which is a 12-digit code, and UPC-E, which is a 6-digit code), Code 39, and Postnet, which is used exclusively for the U.S. mail.

Code 39 is the most popular code used outside the retail industry and the one most likely to be used in EAC. It handles up to 44 characters that can include any of the 225 ASCII characters as well as leading and trailing spaces. The spaces allow two or more bar codes to be scanned as one very long bar code.

Bar codes, which can be printed directly on cards or objects, provide the least expensive, easiest to use system of EAC identification. The software necessary to create bar codes is sold in computer stores and catalogs as well as through industry-specific sources. Readers, which include optical wands, guns, and scanners, are all commonly available.

Unfortunately, although bar codes are convenient for record-keeping, they do not provide an adequate level of security for valuable assets or high security clearance. Bar codes can be easily duplicated by a photocopier or computer.

Because they are easy to create, bar codes can be successfully used for time and attendance reporting and casual EAC. Unlike magnetic encoding, printed bar codes cannot be destroyed by radio frequencies or magnetic field interferences.

Easy duplication via a photocopier or computer scanner can be prevented by placing a special translucent patch over the code. Some patches contain patterns and even logos. No matter what they contain, they blacken the bar code when copied, but still allow optical reading.

The accuracy of bar code reading depends more on the quality and condition of the optical readers used, than the quality of the printed bar code itself. This means that regular reader servicing is required to avoid problems caused by dirty or scratched optical surfaces.

Although bar codes are seldom used alone in EAC, they are often laminated onto more secure credentials, such as Wiegand, magnetic stripe, and proximity cards, to enhance information gathering.

Casual Carriers
Jane Smith
Purchasing Department

Unprotected Bar Code

Casual Carriers
Jane Smith
Purchasing Department

THE LABELS ON THE ROLL ABOVE ARE MADE OUT OF TRANSLUCENT RED MASKING MATERIAL.

TYPICALLY, THIS TAPE IS PLACED ON THE BAR CODE BEFORE THE CARD IS LAMINATED.

Barium Ferrite (BaFe) Cards

Magnetically encoded barium ferrite cards, which were in the forefront of EAC technology during the 1970s, have declined in popularity, although they are still being used. The first barium ferrite card readers were magnetic and mechanical, with many easy-to-damage parts. They worked as follows:

> At setup, a program cartridge, containing a pre-coded array of magnetized spots, was installed in a reader.
>
> Between the program cartridge and the access card insertion area were individual magnets, a movable slider and a metal plate. The slider contained holes through which magnets could fall. The metal plate below the slider stopped the fall of those magnets.
>
> The program cartridge attracted a predetermined array of reader magnets upward and out of the slider holes. The remaining non-attracted magnets stayed in the holes (resting on the plate), thereby jamming the slider in place.
>
> The access card contained magnets positioned to match the pattern of those resting on the metal plate. When this card was inserted, the resting magnets were magnetically repelled (pushed upward), releasing the slider, which slid forward, tripped a microswitch, and released the latch.

At one time, the same code array was used by all the access cards issued in a system. If a card was lost, the program cartridge and the remaining cards had to be reissued. As time went on, additional magnetic spots were added to the main array, forming unique ID numbers that could be interpreted by microprocessors.

The original readers required a great deal of maintenance. As they wore out, many were replaced with competing technologies. Still, as of 1980, there were many barium ferrite cards in circulation, providing the market for a few manufacturers to develop 100% microprocessor-based readers that could interpret other manufacturers' cards as well as produce low-cost proximity-like systems.

Magnetic barium ferrite codes are difficult to duplicate because they are factory-embedded, making them highly secure. They hold up especially well to problems caused by harsh weather and hostile environments, and can be used in mixed technology applications. Like all magnetically encoded cards, however, they can be erased or distorted by strong magnetic fields and tend to wear out over time.

Biometrics

While the security level of credentials is determined by whether or not they can be easily duplicated, unauthorized use can occur when credentials are shared, stolen and/or PINs are exposed. Biometric credentials were developed to defeat this problem by verifying that the unique personal features of the credential user, such as their palm print or eye, match a copy of those features, called a *"template,"* stored in a computer.

Biometrics, which began as an offshoot of the study of genetics and disease, are used when the need for a highly secure identification system offsets the cost of that system.

Various biometric systems have been available for decades, including an attempt by IBM in the 1970s to promote a signature recognition system. Many of these systems, however, were not popular because of high costs, the high rate of verification errors, and verification slowness.

Some biometric templates create extremely large computer files. Fortunately, within the last few years, the prices of powerful computers with massive storage systems have tumbled, along with the costs of new biometric technology.

Today, biometric systems can enroll people in less than a minute and later verify their authenticity in under two seconds with a high degree of accuracy. Some biometric systems can also update individual templates on a daily basis, counteracting the effects of normal aging on recognition, thereby reducing system maintenance.

Most EAC biometric templates are associated with a PIN, which gives the computer quick access to the template. A slower verification mode, which is used in the criminal justice system, requires that the system search through tens of thousands of templates, which might hold fingerprints or blood data, to find a match.

A sample of biometric systems include:

Body odor:

READING PALM PRINTS CHECKS THE LENGTH, WIDTH AND THICKNESS OF THE HAND.

THE ILLUSTRATION SHOWS A HANDKEY, MANUFACTURED BY RECOGNITION SYSTEMS, INC.

Senses odors by using chemical processes that are similar to the processes that take place in the nose and brain.

In early 1995, researchers at Leeds University in England announced that they developed a process that can differentiate between people by their smell. Perfumes do not mask this process because perfumed scents are very different in chemical composition than body odor. Smart card manufacturers hope to eventually embed this technology in their chips in order to

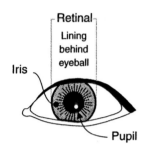

Retinal
Lining
behind
eyeball

Iris

Pupil

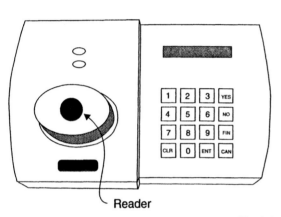

Reader

compete with fingerprinting systems. Before they do that, of course, body odor technology must improve its current identification accuracy of 90%.

Hand and fingerprint identification:

Uses various techniques, among which is a three-dimensional digital image that is captured and measured to create a template. Between 10,000 to 250,000 points of unique information can be encoded in this way.

Eye identification:

There are several ways of using the eye to provide unique identification, two of which follow:

Iris identification measures the iris which, according to product literature distributed by IriScan, can identify 4,000 points in less than three seconds. They claim that iris patterns are fixed at birth and there are no two alike, including those of identical twins.

Retinal scans read the surface behind the eyeball through a low-intensity infrared light that tracks 320 points in the retina and records associated blood-vessel patterns.

Facial recognition

Verifies facial features by comparing a living face scanned by a camera to older images in a database. Because it is easy to change appearance, there are several systems under development that seek to reduce validation time and increase accuracy. This type of technology is far more sophisticated than having a guard check an image database and then determining the similarity between the picture and the living subject.

Multiple biometric patterns:

Assures that a severed body part, such as a finger, cannot be used for falsifying identification by requiring that two different biometric readings be taken at the same time. A blood-oxygen saturation reading taken with a fingerprint scan is an example.

Random voice interrogation:

Assures that a tape recorder cannot be used to bypass a voice-print system, which compares speech patterns. It does this by recording several spoken phrase templates for each person. When identification is requested, the person is asked to recite only one of the prerecorded phrases. Once the phrase is recited, the voice is compared to the appropriate template.

Signature identification:
Measures time and pressure used to create a signature as well as the signature pattern itself.

Voice identification:
Identifies the unique voice characteristics of a freshly spoken phrase to one stored in a template. These comparisons include air pressure and vibrations over the larynx.

Weight measurement:
Although weight is not a biometric measure because it cannot pinpoint specific traits, weight is often used to determine the presence of an individual or thing and consequently, can be used in the authentication processes.

Weight checkpoints are often found in enclosed rooms called "mantraps" as well as around "invisibly" protected objects, such as might be seen in a museum.

While biometrics generally provide a highly accurate verification system, especially when combined with a PIN, users are sometimes concerned about the possibility of physical invasion, harm or discrimination during the credential-reading process. The following considerations describe a few of their concerns:

- A biometric x-ray system, for example, would not be viewed as acceptable because x-rays harm the body with regular use.

- People with certain types of physical disabilities, such as those who have artificial hands or who are blind, might not be able to use the system.

- Blood tests are generally considered too intrusive to do on a regular basis. In addition, there are many regulations governing their use.

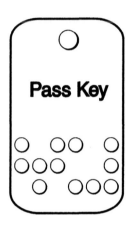

Pass Key

THIS HOLLERITH CARD IS TYPICAL OF THOSE USED IN THE HOTEL INDUSTRY.

Hollerith Cards

Hollerith cards are modeled after cardboard computer cards that were first used in 1890 by the U. S. Census Bureau to automate the national census. These cards featured a uniform pattern of small rectangles arranged in 80 columns, 12 rows high, and held up to 80 alphanumeric characters of information per card.

To encode these original computer cards, keypunch operators punched out selected rectangles, leaving holes that represented values. These cards were then placed in electronic readers which passed current through the holes. The resulting pattern of "on's" and "off's" were electronically translated by a computer into data for number crunching.

Copying the above principle on a simple scale, Hollerith cards also have holes punched in them, but not as many. These thin plastic cards, which can be manufactured in a variety of rectangular sizes, are read optically by passing a light through the holes, or electronically, by allowing metallic brushes to touch contacts exposed through the holes.

Unfortunately, Hollerith cards can be easily duplicated and are only used in low-security applications. Hotels and motels, for example, often use Hollerith cards. When a card is lost, the code can be quickly changed and a new card issued with minimal expense.

Infrared Cards

Light sensitive infrared card technology, also referred to as "transmissive infrared" and "differential optics," appeared in the 1970s and uses bar code principles to encode information.

Embedded in the card is a bar code that is coated in a way that allows predetermined amounts of infrared light to pass through. Electronic infrared sensors detect this internal pattern as reduced energy-level infrared light. The bar code pattern itself cannot be seen by the human eye.

Like bar coded cards, the accurate reading of an infrared card is dependent on the quality and maintenance of its light-sensitive infrared reader. Unlike bar coded cards, these infrared codes cannot be easily duplicated because they are made in a factory and are, therefore, very secure. In addition, they are not subject to erasure by stray magnetic fields as are magnetic stripe, Wiegand, and barium ferrite cards.

Keypads

Keypad devices provide the means to link a PIN with a credential, use a PIN by itself and/or program various devices connected to the system. In all, they are extremely versatile.

Some keypads are part of the locking mechanism. This type of keypad might be programmed to respond to a single PIN that's assigned to everyone, or else, it can be linked to a sophisticated control panel which provides the means to track many codes and time zones.

In most medium to large EAC systems, keypads are linked to powerful control panels and verify cards through use of a PIN. Some keypads even have secret containers in their mountings. These provide a secure storage area in which to place standard keys (for locked cabinets) or other valuables.

THIS IS A TYPICAL KEYPAD AND CARD READER COMBINATION.

Generally, keypads are limited to four- or ten-digit codes, regardless of how many keys appear on the devices themselves. Software, memory and internal circuitry impose these limits, consequently, it is important to examine your PIN requirements before selecting a keypad system. In addition, keypads might not comply with the Americans With Disabilities Act as their location and PIN usage might be difficult for physically and/or mentally challenged people.

Still, keypads are highly durable and are fairly inexpensive to replace. Systems based exclusively on keypads are easy to maintain because they do not require any card or token inventory or related encoding hardware such as is required for magnetic stripe cards.

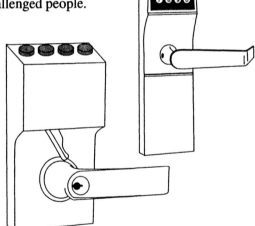

Unfortunately, keypad systems are not highly secure. For one thing, some keypads contain all the wiring necessary to open a door, which means that unauthorized removal can make the lock useless. Another problem is that PINs can be stolen through spying or even casual observation.

The spying issue has been addressed by the ScramblePad, patented by Hirsch Electronics. This keypad reduces the chance of spying success by randomly changing its keytop labels.

INTELLIGENT LOCKSETS LIKE THE ONES SEEN ABOVE CAN FUNCTION INDEPENDENTLY, OR BE LINKED TO A SOPHISTICATED EAC SYSTEM.

On a ScramblePad, the keys labeled 123 might become 976 or 485. Consequently, the finger pattern used to punch in the code 6735 is different with every event. Even if a spy sees the motion, he or she would not know what it stands for. This keypad further reduces spying by shielding its keytop labels and preview window with view-restricting material.

In summary, keypads in combination with card and token systems, play a very important role in EAC and are in common use.

Magnetic Stripe Cards

Most people have seen and used a magnetic stripe card of some type. These cards are the most widely used cards in the world and proliferate as bank, credit and, of course, EAC cards. They are inexpensive, can carry alphanumeric information, are quickly produced and can be encoded at the user's site.

Each card contains a black plastic stripe of magnetically-sensitive oxide, which is the same material used to make audio/video tapes. Unlike tapes, however, magnetic stripe cards are subjected to frequent rubbing and bending. Despite their lack of protective housing, their ability to retain magnetically encoded information is quite good.

The risk of magnetic erasure, however, is always a problem. Their resistance to erasure is known as their *coercive force rating*.

Coercive force ratings indicate the strength of a magnetic force required to erase magnetic material. A card with a low coercivity rating, therefore, is fairly easy to erase, and a high coercivity rating means that the card is more protected from stray magnetic fields. Needless to say, disposable cardboard cards are more likely to have a lower coercivity rating than more permanent plastic varieties.

According to the American National Standards Institute, Inc., magnetic stripes must contain four tracks available for encoding, however, specific encoding standards have only been defined for tracks one and two:

- *Track one:* Stores up to 79 alphanumeric characters (210 bits per inch). This information might include the user's name and maybe a title.

- *Track two:* Stores up to 75 bits per inch, with 40 numeric characters. This is the track most commonly used for access control codes.

- *Track three:* This track can contain a facility code (also known as a *water mark*), which is described later in this section. Access to track three requires a special dual-head reader.

Magnetic stripe cards store more characters of information than associated with bar coding or magnetic particle embedding and are far easier to customize.

All encoding can be done at the end user's facilities by manual or automatic equipment. Manual encoders, of course, are more cost-effective for organizations that issue only a few cards. Automatic encoders speed up the process for issuing multiple-cards, plus provide more control features. These include assigning sequential issue numbers and printing images on the cards, in addition to the encoding process itself.

Unfortunately, with the right equipment, unauthorized duplication of magnetic stripe information is possible, rendering their security somewhat low. It is common, however, to see magnetic stripes on mixed-technology cards, which increases their security level. One very new development, for example, encodes highly secure, machine-readable, hologram patterns and a magnetic stripe on a single card. The combined use of PINs with magnetic stripe cards, of course, is well known.

Magnetic stripe information is read by means of a swipe (moving the magnetic stripe along a track that passes a reader head) or insertion. Swipe readers, with their exposed reader heads, should only be used in environmentally clean areas. Insertion readers are less affected by environmental dust and are suitable for outside installations.

Motorized insertion readers regulate the speed at which the card passes the reader head and may increase reading accuracy. Like tape recorders and VCR's, however, the quality of the information transfer under any circumstance is largely dependent on the strength of the magnetic properties in the magnetic stripe and the cleanness and orientation of the card reader head.

Magnetic stripe cards can be individualized by photos and/or bar codes through lamination, printing, or dye sublimation. Care must be taken to make sure that bulky lamination does not jam up card travel in the reader.

Facility Codes and Water Marks: For a variety of reasons, some systems restrict the number of characters that can be used in a card code, thereby limiting the total number of codes that can be issued. Others restrict the number of codes a control panel can interpret before polling a main computer.

To increase the number of card codes available, a special code, called a *facility code* or *water mark*, is permanently fixed in Track Three by the card manufacturer. This is done through a proprietary system that positions magnetic oxide particles on Track Three via a wet slurry. When the slurry dries, the information is secure.

The result of applying a facility code is that a range of card codes, such as from 0001 to 9,999, can be duplicated. In the following example, a 10,000 card code range is expanded to approximately 30,000 possibilities.

> Facility Code 1002 - range 1-9,999
> Facility Code 1003 - duplicate
> Facility Code 1004 - duplicate

In addition to increasing the number of available codes, facility codes can be used to detect tampering and unauthorized card duplication.

Mixed-Technology

The ideal credential should be capable of combining a variety of technologies, including proximity, magnetic stripe, microprocessor (smart card), Wiegand, infrared, and keypad. In addition, users should be able to inexpensively apply customized designs, photos, and/or bar codes to cards for further individualization.

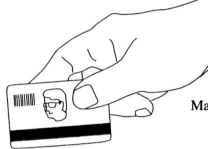

One advantage of using mixed-technology is that a single card can be read by different types of readers. This makes retrofitting (updating) existing card systems more cost effective because it doesn't require replacement of hardware or wiring. Another advantage is that it reduces the number of cards a person needs to carry.

Many universities are taking advantage of mixed-technology card systems:

> *Example:* In one college, a student photo ID card uses proximity technology to unlock dorms, bar codes to track library books, and a magnetic stripe with PIN to access the debit system used by the cafeteria and ticket agents.

The three most common credential technologies are magnetic stripe, Wiegand, and proximity. Wiegand and proximity cards offer an exceptionally high degree of security. Magnetic stripe cards carry a great deal of information and are easily encoded. Proximity cards, which do not touch their readers, improve traffic flow and reduce reader maintenance costs.

Combining the three technologies mentioned above into a single credential requires:

1. A standard card size (CR-80 or CR-60 as mentioned earlier in this chapter).

2. A card thin enough (.03") to fit through a magnetic stripe swipe or insertion reader.

3. Sufficient voltage to drive Wiegand and/or proximity systems.

Other combinations are possible, too. Biometrics, for example, often require a huge computer file (template) for each authorized person. Verification might take an excessive amount of time if the biometric reader has to check against templates held in a distant computer. Smart cards, however, can easily hold these large files. This allows the use of a biometric system for personal identification without being tied to a distant database, thus avoiding problems associated with slow or poor telecommunication connections.

Optical Cards

Optical cards are very new and are not widely used for EAC. This type of card was first developed by Canon U.S.A., Inc., and can store between 3.42 to 4.20 Mbytes of data on the size of a credit card. The amount of storage space it contains depends on the sensitivity of the card reader itself.

The benefit to such a credential is that it can carry an enormous amount of information, thus reducing the telecommunications time a reader might require to seek details from a distant computer. The disadvantage is that this information must be entered on the card at the factory.

Optical cards are created by a solid-state, high intensity, laser beam that burns tiny pits on the card's surface. To read this data, a low intensity laser beam directs light on the pits, the reflection of which varies in accordance with the data that was initially etched. The Cannon system writes information on 2,500 tracts, some with multiple sectors, using the same write-once-read-many-times (WORM) techniques as for creating CD disks.

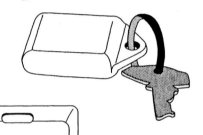

Touchless
Corporation

THESE PROXIMITY TOKENS CONTAIN A MAGNETIC COIL, MEMORY CHIP AND A BATTERY.

SLIM CARDS CONTAIN EVERYTHING BUT THE BATTERY.

THE BRACELET TOKEN IS COMMONLY USED IN HOSPITALS AND KEYCHAIN TOKENS FOR GARAGE APPLICATIONS.

THE FLAT PANEL TOKEN CAN BE KEPT IN A CAR OR ELSE ATTACHED TO A CLIP FOR WEARING.

THESE ILLUSTRATIONS ARE BASED ON TOKENS MANUFACTURED BY COTAG INTERNATIONAL.

Proximity Cards and Tokens

Popular proximity cards or tokens do not need to touch a reader in order to validate a transaction. Their radio-wave transmission technology is highly secure and their readers can be hidden behind walls, in clean, maintenance-free, vandal-proof locations. Best, because proximity readers don't require contact, they speed the flow of human and vehicular traffic through check points. These cards and tokens can remain in pockets, purses, or even on the front seats of vehicles and still be read by the system.

Although proximity cards were pioneered in the 1970s by Schlage Electronics (which was later purchased by Westinghouse Security Electronics), they weren't used extensively until 1992, when two big vendors, Hughes Identification Devices and Texas Instruments, entered the radio frequency identification industry (RFID). Texas Instruments concentrated on industrial applications, while Hughes Identification Devices developed EAC applications. In 1993, another RFID company, Indala, became a subsidiary of Motorola, bringing competitive forces to bear. By 1994, the price of proximity cards dropped and is now in line with other technologies.

Electrical power is always a big EAC concern and, prior to 1993, a typical proximity reader drew a great deal of current (400 mA at 12 volts). Today, readers can operate with current as low as 40 mA at 5 volts, which is the same range used for Wiegand and magnetic stripe readers.

Each proximity card contains a coil of wire that acts as both the receiving and transmitting antenna and a small, integrated circuit that is programmed with a unique ID code. The cards are powered by the voltage generated from a reader's magnetic field in relation to the card's antenna coil. Once energized by a reader, the card transmits its ID information. Transmission is so fast that access verification takes place in less than a quarter second.

There are two types of proximity cards or tokens:

- **Active:**

    Has a range measured between touch to 100 feet. Its transmission is powered by a small, lithium battery. Due to battery thickness, active proximity carriers are manufactured as tokens or thick plastic containers that look somewhat like cards. The battery loses power over time and requires systematic replacement, although its average life is from five to seven years.

● **Passive:**

Has a range measured between touch to 30 inches. Its transmission is powered by magnetic properties embedded in a very thin, maintenance-free card. The technological trend is to extend its transmission distance through the use of space technology that was originally developed to receive faint signals from distant stars.

Proximity technology is secure, reasonably priced and becoming increasingly used in mixed-technology cards. Readers have been miniaturized to fit into spaces less than 1.75" square and are getting smaller every day. Being convenient, they comply very well with regulations defined by the Americans With Disabilities Act.

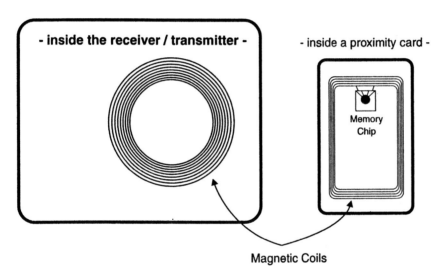

- inside the receiver / transmitter -

- inside a proximity card -

Memory Chip

Magnetic Coils

How Proximity Works

1. A receiver/transmitter (R/T) is either buried in a wall or contained in a slim cabinet hung on a wall.

2. The magnetic coil in the R/T excites the magnetic coil in a card when that card is in range. This range is extended when the card or token contains a battery.

3. Once it is excited, the magnetic coil in the card generates a crisp, magnetic pattern that represents a code contained in its memory chip.

4. The R/T receives the magnetic pattern and responds by amplifying and transmitting it to the processor — a control panel or other unit.

Smart Cards

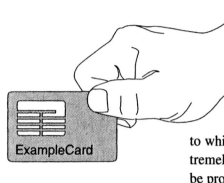

ExampleCard

Although smart cards are currently uncommon in America, that may soon be changing. European companies (telephone systems and banking) have been using smart cards extensively since the early 1990s.

A smart card is essentially a credit-card sized computer that was invented over 20 years ago. Embedded in the card is a microprocessor with memory that can be read and, more importantly, written to which can store a significant amount of information. Counterfeiting is extremely difficult because the chip is buried in plastic. In addition, the chip can be programmed to generate its own passwords and codes, including sophisticated encryption functions.

SMART CARDS LOOK LIKE COMMON CHARGE CARDS AND CAN EVEN HAVE A MAGNETIC STRIPE, BAR CODE AND/ OR ID PHOTO PRESENT.

YOU CAN IDENTIFY A SMART CARD BY A METALLIC DESIGN SIMILAR TO THE WHITE ONE SEEN ON THE CARD ABOVE. THIS DESIGN AREA HOLDS THE COMPUTER CIRCUITRY.

The trend today is to embed a significant amount of information in a card in order to reduce time-consuming access to distant computers. Biometrics, for example, require large computer data files (called *"templates"*) to store complex physiological patterns. By keeping that information on a smart card, identification time is greatly reduced and worldwide check-in sites (such as used in the military) are not subject to long-distance communication problems.

There are three types of smart cards:

- **Memory-only:**

  Has less than 400 bits of memory and are often used for disposable "prepaid card systems."

- **Memory circuits with some hard-wired security logic:**

  Contains between 1K to 4K of memory and can be erased and rewritten. These are designed to allow encryption and PIN comparisons.

- **Full-fledged microcomputers:**

  Contains a complete computer system with an operating system and the ability to be programmed to meet a wide range of applications. The computer system includes a processor, nonvolatile read/write memory of 1K to 8K, a small amount of random access memory, and read-only memory which contains the operating system and the place where security functions are hidden.

Although a smart card looks similar to a standard credit card, it differs by having five to eight metallic contacts displayed on its surface. These contacts connect directly to a computer terminal when the card is inserted. To make sure that contact is good, smart cards must be stored flat at all times and maintenance is needed to make sure that terminal readers are clean.

To increase the potential markets for smart cards, information held in the chip can now be transferred through proximity technology. Although contactless and radio communication methods have not yet been standardized, systems are available.

New uses for smart cards are being invented every day. Several states, for example, are replacing food stamps and other voucher systems with smart cards. These cards reduce paperwork and theft, while increasing reliability and ease of benefit transfer. Hospitals are also using smart cards for EAC as well as for sharing patient records. Updating a smart card from a single computer source, as opposed to transferring information from numerous charts and records, improves communications, increases accuracy, and decreases costs associated with paperwork.

Wiegand Cards

Wiegand cards are based on patented technology held by the Sensor Engineering Corporation that embeds an array of magnetic wires in a card that is very difficult to duplicate.

Wiegand technology combines several patented processes and a special metal alloy to create unique magnetic properties not found in common ferromagnetic (iron) wires. Through manipulation and heat-treatment, the core of a Wiegand wire acquires a different magnetic property than its shell. This results in a condition called *bistable magnetic action*.

Bistable magnetic action creates an electrical pulse:

- When first subjected to a strong magnetic field, the wire has a magnetic north and possesses a unified external magnetic field.

- When the wire is then subjected to a weaker magnetic field that has a south orientation, the wire's core switches its polarity to the south, while its exterior shell remains north. This causes the wire's external magnetic field to collapse.

- When subjected to the original strong magnetic field again, the core reverses its polarity to match that of its exterior shell. This change in polarity creates a crisp, discrete electrical pulse.

The only energy input required to create the electrical pulse is the bistable action of the wire in relationship to variations in magnetic fields produced by the reader. Although the pulse is regarded as analog, it is so crisp that it can be read as a digital output.

Every Wiegand wire segment embedded in a card represents a magnetized pulse-generating "bit." Up to 56 bits are allowed per card, although in reality, not all the bits are required.

These bits are arranged in two parallel rows. Bits in the top row are referred to as *zero bits*, and in the bottom, *one bits*. The placement of these bits in relation to one another generates a binary pattern that represents a unique code.

Wiegand readers have two reading heads, one for each row, that read the electrical pulses generated by the bistable magnetic wires.

Manufacturing Processes: All Wiegand wire is tested three times before it is cut into .33" strips and placed on vinyl adhesive tape in a pattern determined by computer-controlled machinery.

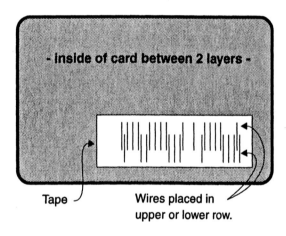

- inside of card between 2 layers -

Tape

Wires placed in upper or lower row.

For each order, the wire-encoded tape is spooled onto a tape feeder, then fed to a tape-cutting-and-placing machine. This machine automatically cuts and places 12 strips of tape in appropriate locations on vinyl sheets (which are eventually cut into 12 cards), continuing with new sheets until the order is complete.

Once a vinyl sheet is prepared by the encoded tape and topped with artwork and lamination, it is pressure-temperature sealed, die cut, and inspected. Depending on the order, some cards also receive a special hot stamped card number. Before being shipped to the customer, the cards are inspected. Cards not meeting the inspection criteria are discarded and remanufactured.

Changes in Technology: The original Wiegand cards were somewhat thick. New credit-card thin sizes now allow Wiegand technology to be combined with other popular technologies, such as magnetic stripes. In addition, other manufacturers are developing proximity readers that can pick up Wiegand's low voltage power output (5 volt, 25 mA), while Sensor Engineering Corporation itself has also developed proximity technology.

Wiegand cards are regarded as very durable, secure, and reasonably priced. Unfortunately, because Wiegand cards are crafted at the factory, they take longer to manufacturer than other cards, forcing some EAC system managers to use other technologies because of lead-time considerations.

More Information on Credential Technology

To further your understanding about credential technology, attend trade shows, such as sponsored by the American Society of Industrial Security (ASIS) or the annual Access Control Conference/Expo sponsored by *Access Control Magazine*. There you'll see examples of all varieties of credential carriers as well as have the option of attending seminars on related subjects.

See the *Appendix* for more information.

Chapter Review - Credentials

When presenting a card or physical feature to a reader or scanner, the . . .

card is *authenticated,* and a

physical feature (such as a finger print) is *verified.*

Three most commonly used credential technologies are:

1. Magnetic stripe

2. Wiegand

3. Proximity

A credential reader is similar in concept to a disk drive on a personal computer because . . .

it reads the information found on credentials and passes that information to a computer for final processing.

Although readers can be used for a variety of computerized applications, EAC readers must have . . .

tamper-resistant mountings.

Three most common ways to apply an image to a card are:

1. Lamination, using a regular photo or Polaroid

2. Dye sublimation, using a digital image captured by a computer

3. Digital storage in a smart card chip, using a digital image captured by a computer

Two standard card sizes are:

1. CR-80 (2.125" tall by 3.375" wide by .03" thick)
2. CR-60 (2.375" tall by 3.25" wide by .03" thick)

The primary difference between the two standard card sizes is that . . .

the CR-80 is the most universal and can fit in both insertion and swipe readers. The CR-60 can fit in all swipe readers, but requires a special insertion reader.

Twelve EAC credential types are:

1. Bar coded cards and objects
2. Barium ferrite cards
3. Biometrics
4. Hollerith
5. Infrared cards
6. Keypads
7. Magnetic stripe cards
8. Mixed-technology
9. Optical cards
10. Proximity cards and tokens
11. Smart cards
12. Wiegand cards

Bar coded credentials are . . .

> easy to create, however, are not very secure because they are very easy to duplicate. Bar coding is often used in conjunction with other credential technologies.

Barium ferrite cards are . . .

> considered to be somewhat obsolete in light of newer technologies, however, they cannot be easily duplicated and are very secure. Encoding must be done at the factory rather than at the user's site. They can be erased or distorted by strong magnetic fields and can wear out over time.

Biometrics are . . .

> used when an exceptionally high level of security is required and work by associating unique physical characteristics with personal identity.

Hollerith cards are . . .

> based on holes punched in cards through which a pattern of light or electrical current can pass. Although the basic hole pattern can be easily changed as needed, once issued, the cards themselves are easy to duplicate and are not regarded as being highly secure.

Infrared cards use . . .

> bar code technology, however, the bar code is embedded in the card itself, is not visible to the human eye, and cannot be duplicated. Encoding must be done at the factory rather than at the user's site. They are not affected by magnetic fields and can have a very long life.

Keypads provide . . .

> a miniature keyboard. They are durable and reasonable in price. User PINs can be stolen if care is not taken to hide keypad entry from the view of casual onlookers.

Magnetic stripe cards are . . .

> the most widely used cards in the world. They are inexpensive, can be encoded at the user's site, and carry an adequate amount of alphanumeric information. They are affected by strong magnetic fields and wear out depending on frequency of use. Magnetic stripes are often used with other technologies, such as keypads.

Coercive force ratings indicate the strength of a magnetic force required to erase magnetic material:

1. Low ratings mean that the magnetic material is fairly easy to erase.

2. High ratings means that the material is more protected from stray magnetic fields.

Mixed-technology cards provide . . .

> several technologies on one card. This makes retrofitting older systems more cost effective, allows more information to be contained on a single card, and reduces the number of cards a person might need in a single facility (such as in a college).

Optical cards are . . .

> the latest in technology, are very expensive, must be encoded at the factory, however, are very secure. Similar to CD-ROM disks, they can hold a great deal of information.

Proximity cards and tokens are . . .

> dropping in price and growing in popularity. They transmit information via unique radio waves to a reader that can be "invisibly" buried in a wall. Some proximity cards can be customized at the user's site.

Two types of proximity cards or tokens are:

1. *Active:* depending on battery power, transmits information over distances of 1 foot to 100 feet. They often take the form of tokens, which form a case large enough to hold the lithium battery.

2. *Passive:* transmits information between 1 to 30 inches. They are usually very thin cards similar to magnetic stripe cards.

Smart cards have . . .

a computer embedded within the card which can carry an enormous amount of user-created information as well as sophisticated encryption systems.

Three types of smart cards are:

1. Memory-only

2. Memory circuits with some hard-wired security logic

3. Full-fledged microcomputers

Wiegand cards are . . .

based on highly secure, patented technology that transforms a metal alloy wire into having *bistable magnetic action*. This action causes the wire to quickly change polarities based on magnetic attraction and produces a discrete electrical pulse. While popular and regarded as very secure, the cards must be encoded in a factory and require a longer lead time in comparison to other technologies.

QUESTIONS

1. Name 12 types of EAC credentials.

2. Why should a reader mounting be tamper-resistant?

3. What is the critical difference between the CR-80 and CR-60 magnetic stripe card size?

4. The majority of EAC systems use what three types of technologies?

5. What does a coercive force rating mean when rating a magnetic stripe and why is it important?

Chapter 4

Barriers

Doors, Gates, Turnstiles, and Electric Locks

It is easy to design passages that are difficult to enter. We can, for example, install heavy barriers, multiple locks and bolts, obstacles, and hidden traps. Fortunately, where the general public is concerned, safety rules require that no matter how difficult it may be to get in, people must be able to leave easily.

Barriers, doors, and locks used in public buildings are regulated by extensive legislation and local building codes. Engineers, architects, builders, consultants, locksmiths, police, fire, and regulatory agencies are all involved with designing, selecting, and installing the barriers used to control access. If any of these professionals fail to do the right thing, lawyers, insurance companies, and the courts have the ultimate say.

Security professionals, especially those in highly secure situations, specialize in pinpointing, then eliminating, weaknesses in facilities. Many electronic access control (EAC) managers, however, must accept physical weaknesses in their facilities (such as glass windows and doors). They counter this by increasing their watchfulness.

This chapter, then, provides an overview of the complex subject of barriers, including electric locks, but it is not exhaustive. For further information about these devices, check the appendix for names of associations, magazines, and rating laboratories. Two magazines, in particular, are worth investigating:

> *Locksmith Ledger International*, The Locksmith Publishing Corp.,
> 708 / 692-5940

>> Monthly. Illustrates concepts related to locks, doors, and related security equipment.

> *SDM Field Guide*, Security Distributing and Marketing,
> 800 / 662-7776

>> Quarterly. Illustrates security-related planning concepts including wiring guides, EAC, CCTV, and sensor placement. Includes highly informative, well illustrated "tests of knowledge and skill."

To get more out of this chapter, observe the barriers and locks you see in your daily life. Especially notice doors in shopping centers and medical complexes, fences around industrial areas, turnstiles, revolving doors, gates in parking lots, etc.

Exterior Barriers

In ancient times, thick stone walls secured the perimeters of villages and forts. As you read in Chapter 2, these stone walls were surrounded by clear viewing areas (glacis), walkways providing guard patrol areas and galleries serving as lookout points.

With the reduction of clan wars and the increase of peace, today the majority of our buildings feature minimum barriers, thin walls and many glass windows. Even buildings at risk are relatively open and inviting. In general, we have replaced brute force and heavy barriers with *surveillance equipment*, and more important, intelligent friendliness.

Electronic access control is associated with some barriers because it serves to open gates, regulate traffic and, with the aid of strategically located sensing devices, report problems. Sensors, woven through fencing barriers, for example, report problems when that fencing is stressed.

To be regarded as a barrier, a structure or natural area must perform at least three functions:

1. **Define:** provide clear boundary markings of the area to be protected.
2. **Delay:** delay unwanted traffic, but not necessarily stop it.
3. **Direct:** direct traffic to proper entrances.

All barriers are designed to discourage three types of penetration. The first is penetration by *accident*. It is just about impossible, for example, for someone to walk through a securely locked door, but easy if the door is left open. Consequently, warning signs are usually posted by barriers to reduce accidental access. The last two types of penetration are by *force* and by *stealth*.

The problem with penetration by force is that it ruins property. On the other hand, forced entry can quickly alert the authorities, plus leave evidence that helps to track the intruders. Penetration by stealth is very serious because the authorities may never know that a crime was committed.

Security designers are sometimes more concerned about combatting penetration by the use of force and stealth than they are accidental penetration. The design of Super Max, an ultrahigh security United States prison, located in Denver, Colorado, is an example:

> There are seven layers of steel and poured cement between the outside walls and the cells. The yard is bounded by 12-foot fences, topped with razor wire and overseen by six towers manned with armed guards. Guy wires are strung across the yard to prevent helicopter escapes and the entire facility is wired with state-of-the-art sensor technology.

Defeating penetration is often difficult.

Nature can nullify the purpose of exterior barriers, especially when these barriers are not rigorously monitored. Here are some examples:

- Drifting snow makes barriers easy-to-mount by piling up along the sides.
- Deep snow is easy to trench, which lets invaders sneak around unseen.
- Bushes, trees, and other vegetation that grow next to barriers provide natural ladders, hiding places, and areas to burn.
- Rain makes it easy to tunnel into the ground beneath the barriers.
- Heavy fog and/or driving rain hides unauthorized access attempts from view.

Structural barriers are also a problem:

It takes as little as 45 seconds for a man to batter a hole through an 8-inch, mortar-filled concrete block wall with a ten pound sledge hammer. Add another minute to that time to punch through a 5-inch, mortar-filled concrete block wall containing 1/2-inch steel reinforcing rods.

It takes a split second to break glass and between one and 2.25 minutes to punch through 9/16-inch security glass. Common doors can be breached in under a minute.

Even though unauthorized penetration is always a concern, the majority of EAC managers work in areas where the public has great accessibility. In these situations, more attention is given to watchfulness and traffic control than to building and guarding thick barricades.

As you can see from the statistics above, it does not take much for a willful person to break through walls and doors. Unattended and neglected property is always most at risk. The lack of physical might, however, does not mean that access control is not taken seriously!

Vehicle Barriers

Vehicle barriers delay unauthorized access, but do not necessarily prevent it. Many gates, in fact, are quite flimsy and, unless attended by guards, let pedestrians easily walk around.

Making sure the flow of traffic goes in the right direction, however, is important. Borrowing on ancient ideas for castle defense, some "wrong way" vehicle entrances hide spikes or other objects underground. When someone drives over them in the wrong direction, the spikes pop up and their tires are destroyed.

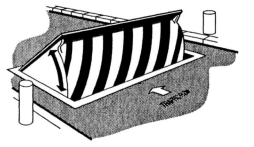

Another type of pop-up barrier is elevated after-hours and then lowered during normal traffic. These can appear in the middle of driveways or in conjunction with lowered gates.

Lightweight gates stop cars from ploughing through but are not meant as a threat to the driver. In northern climates, where ice-slicks can impose a hazard, you would not want a gate so strong that would kill the driver if his or her vehicle slipped out of control. Likewise, you would not want to decapitate a driver if his or her vehicle was hit from behind.

Heavy motorized gates are often left open during peak traffic periods in order to reduce wear on their mechanisms or to simply save time. In more secure areas, guards posted in nearby stations observe vehicle traffic.

Vehicle surveillance devices include closed circuit TV (CCTV), lighting, two-way speaker systems and a variety of sensing devices. Some of those sensing devices detect the presence of vehicles, while others sense when pressure has been applied to the fencing material itself.

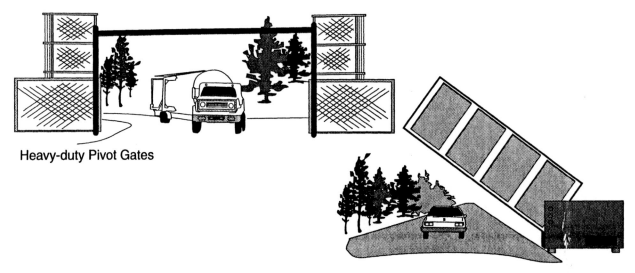

Heavy-duty Pivot Gates

Card and Proximity Vehicle Control

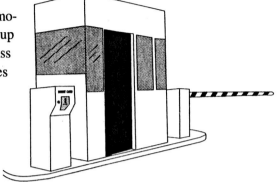

Card readers used at many EAC vehicle gates require every motorist to stop, roll down a window, insert a card, then roll up the window again. All this can be irritating to drivers who pass the same checkpoint daily and the slowdown often causes traffic backups.

Fortunately, the railroad and trucking industries combatted this problem by pioneering continuous-motion vehicular access control. Before employing this technology, freight haulers were required to stop, process forms by hand, wait for authorization, and then move on. With this technology, the haulers are now able to roll through checkpoints without stopping, wasting time, or energy.

It works as follows:

> Each authorized vehicle contains a proximity token or bar code. These vehicles can be granted access by driving through an EAC checkpoint at speeds of up to 30 MPH without having to stop.

> The proximity or bar code reader, located on a tower, bridge, or on the side of a building, can sense tokens or bar codes within 30 feet and grant access in less than 1/10th of a second. Nonconforming vehicles, of course, trip alarms, or are directed to more secure gates for standard check-through procedures.

This technology has been adapted to automobiles and eliminates stopping at checkpoints. It now controls commuter traffic at international border crossings, toll roads and large parking lots.

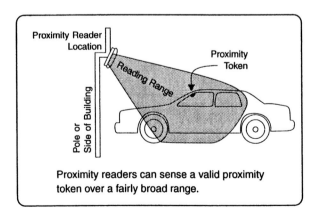

Proximity readers can sense a valid proximity token over a fairly broad range.

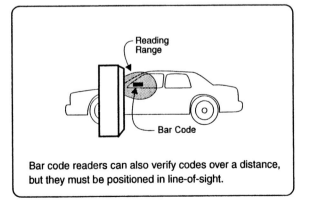

Bar code readers can also verify codes over a distance, but they must be positioned in line-of-sight.

Chapter 4

Doors

Function, governmental regulations, appearance needs, and cost all determine a door's style and materials. Beyond these factors, doors controlled by EAC have the following in common:

- A door closing mechanism

- An electronically or magnetically activated lock

- Sensors (switches) that determine whether or not the door is properly closed

- Computerized control either in the locking device itself, or in a nearby, hidden control panel.

These points are discussed in the remaining portions of this chapter.

EAC requires that doorways, walls, and ceilings have a power source nearby and, in most cases, have adequate conduit and ducts in the walls or ceilings to hold electrical wiring. In areas where placing wire is difficult or prohibitively expensive, such as in an old elevator shaft, wireless EAC is substituted.

EAC systems also require wall or ceiling cavities large enough to contain control panels or wiring closets.

Door Closers

Mechanical door closers are as important to an EAC system as the electronics that power the locks. Door closers are spring-activated with tension strong enough to pull doors completely shut after use, yet not so strong that it makes opening the door a struggle, or warps the door during normal use. The mechanism attached to the spring that guides the door shut is called the *arm*.

Door closers fall into two main categories: concealed and surface mounted, a sample of which is seen on the next page.

Concealed closers are usually used on doors designed for a clean, "no-hardware" look because the arm is hidden from view. They can be difficult to adjust and service, however, because they are embedded into the top or bottom of the frame and door itself, requiring the door and frame to be perfectly balanced. Hardware replacement is usually manufacturer-specific and requires exacting specifications.

Surface mounted door closers are the most popular and fall into three main categories:

1. Regular-arm mounted
2. Top-jamb mounted
3. Parallel mounted

The most popular are the regular-arm and top-jamb styles, the latter of which is simply the regular-arm style installed upside down.

- The *regular-arm* and *top-jamb* door closers can stand the greatest deviation in door play. They are usually installed on the interior-side to reduce tampering, reduce weather damage such as rusting, and enhance the exterior appearance of the door.

- The *parallel style* is less popular. The arm on this closer slides parallel to the door, rather than perpendicular to it. Unfortunately, it is difficult to service because it requires a very well-balanced door. This type of closer is usually used when a jamb mounted closer *must* be installed on the weather-side of a door. It is thought to be more weather-resistant because its arm does not stick out and it can be shielded by a roof of some sort.

The series of illustrations on the next page show where door closers are commonly positioned on doors. Your understanding of these mechanisms can be greatly enhanced by observing the doors in public and private buildings.

Door Closers

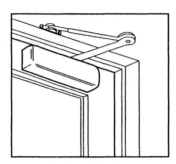

Surface Mount
Regular-Arm

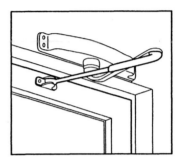

Surface Mount
Top-Jamb

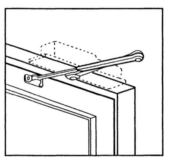

Overhead Concealed Mount
Exposed Arms

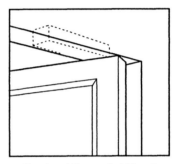

Overhead Concealed Mount
Pivoted Door

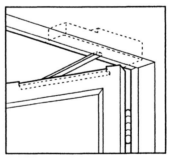

Overhead Concealed Mount
Hinged Door

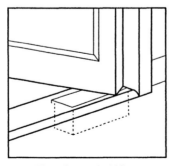

Floor Concealed Mount
Pivoted Door

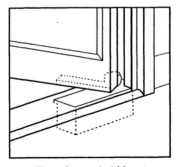

Floor Concealed Mount
Offset Pivoted Door

Electronic and Electromagnetic Locks

The four most common types of locks used in EAC systems are the magnetic lock, the electric strike lock, the electric lockset, and the electric dead bolt.

The strength of any lock is determined by the way in which it can stop unauthorized entry that uses cleverness or force. Electronic or electromagnetic locks, therefore, must be strong enough to guard against:

- Picking (where parts are manipulated)

- Drilling (which destroys the device)

- Electronic or magnetic trickery (which includes the use of unauthorized credentials and the manipulation of the power supply).

EAC electronic and electromagnetic locks are regarded as being either fail-safe or fail-secure and both have an important role in overall security:

Fail-safe:

The lock is *unlocked* when the power is off. This type of lock is usually used on a fire door. In the event of a fire, the locks can be released through the fire system or, if the power system fails, they unlock automatically.

Fail-secure:

The lock *remains locked* when the power is off. Power is required to unlock this type of lock and is usually used for normal locking situations.

The following pages provide an overview of the lock types.

Magnetic Locks

Magnetic locks secure doors through magnetic force and are *always* fail-safe devices. They are ideal for high-frequency access control usage because they are totally free of moving parts, which reduces wear and tear.

Every magnetic lock consists of two components:

- Electromagnet
- Strike plate

The electromagnet is installed on the door frame and the strike plate on the door itself. When energized, the electromagnet attracts the strike plate with a holding force ranging between 500 lbs. to 3,000 lbs.

All EAC systems require that some form of sensor reports whether a door is open or closed. Conveniently, many magnetic locks have that sensor built in, eliminating the necessity for a secondary sensor or switch.

The two basic magnetic lock styles are called:

- Direct hold, which is surface mounted on the secure-side of the door frame and door.
- Shear (also called *concealed*), which is completely embedded within the door frame and the door itself.

The large, *direct hold* magnetic lock is ideal for use on poorly fitted doors and unframed glass doors because the two lock parts can be installed in rough proximity to each other. When energized, the electromagnet positioned on the frame attracts the strike plate on the door flush to its surface. This strong attraction doesn't require perfect horizontal or vertical alignment between the parts.

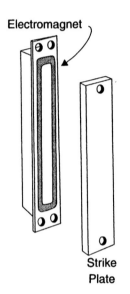

Electromagnet

Strike Plate

Direct Hold Magnetic Lock
with power cord exposed

Shear (hidden) Magnetic Lock

Smaller *shear* magnetic locks, which are less than door thickness wide, are totally invisible to the eye when the door is closed. They are used when design and aesthetic considerations dictate that the lock be completely hidden. Concealing reduces the potential for tampering because the electrical wiring is completely enclosed within the door frame.

The narrow surfaces on the shear electromagnet and the strike plate require precise alignment. A small bracket is often used on the frame to stop door travel so that these surfaces line up.

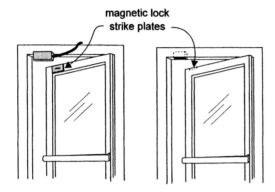

ANSI standards have defined three grades of magnetic locks. Grade one, which holds 1,500 pounds, is designed for medium security. Grade two, at 1,000 pounds, is for light security, and grade three, 500 pounds, simply holds a door shut. Most 180 pound men can force open a door equipped with an 850 pound magnetic lock.

As the holding attraction increases to 2,000 or more pounds, a magnetic lock will stay joined even when the force of a blow is strong enough to shatter the door it secures. Consequently, in addition to the strength of the lock itself, the material strength of the door, frame, and wall must also be considered when planning a high security door.

Electric Strike Lock

The electric strike lock is the most popular EAC locking device on the market and can be set up as either fail-safe or fail-secure. Its popularity stems from the fact that it comes in a wide variety of sizes and can replace existing mechanical locks without a great deal of difficulty. The strike, which is the electrically controlled portion of the lock mechanism, is mounted in a door frame (jamb) and does not require wiring through the door itself.

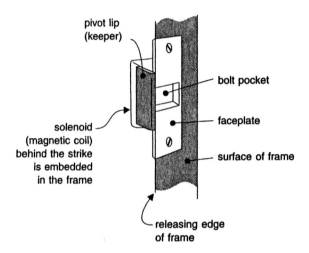

The electronic strike contains a bolt pocket, which is the indent that holds the protruding latch bolt or dead bolt secure in the frame. To open, the strike rotates away from the pocket, providing a path for the bolt to escape. This rotating side is called a *pivoting lip* or *keeper*.

The latch bolt or dead bolt housing itself is mortised (embedded) in the door.

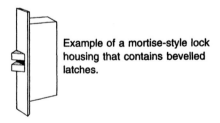

Example of a mortise-style lock housing that contains bevelled latches.

Latch Bolt: The latch is a spring-loaded, beveled bolt. When the door closes, the beveled-side of the bolt slides over the strike, allowing the bolt to retract and then expand again in the bolt pocket once the door is fully shut.

Dead Bolt: The dead bolt is a solid metal rod or rectangularly-shaped bolt that has only two possible positions: protruding or retracted. The protruding bolt enters or escapes the bolt pocket in the frame only when the pivoting lip of the electric strike is rotated away from the frame.

The solenoid (magnetic coil) that activates the strike receives low AC or DC current through a power cord hidden in the frame. A soft buzzing noise can often be heard when AC current is used. This is caused by the vibrations of the alternating current pushing and pulling the solenoid 60 times per second.

Electric strikes and their related latch bolts come in a variety of styles suitable for installation on wood and metal frames. Each frame type, however, poses its own demands. A few of the many things to consider include:

- Wood frames can be weakened from the hollowing out required for installation of the electric strike and need additional anchors or brackets to protect the lock itself against forced-entry attempts.

- Tubular aluminum frames might be too shallow to accept an electric strike assembly.

- Hollow metal frames might be too weak to resist a forced entry, or else were filled with cement or plaster when installed, prohibiting the installation of the electric strike at a later date.

Electric Lockset

The electric lockset is very similar to a mechanical lockset and is available in cylindrical and mortise styles. The difference is that an electric solenoid (magnetic coil) replaces the mechanical action provided by a standard key. In addition, only the electric lock has fail-safe or fail-secure operational modes.

- *Cylindrical Lockset:* These are characterized by a door knob or handle on each side of the door which are joined by a cylinder that controls the locking mechanism.

- *Mortise-style Lockset:* These are characterized by a lock, which is housed in a rectangular metal container, that is embedded at the edge of the door and is often enclosed within the door's thickness.

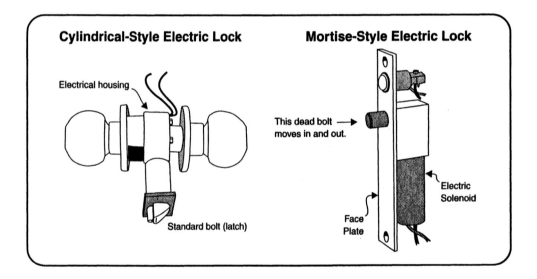

Electric power is brought to the lock by threading wire from the frame through the door. Electric hinges (or pivots) completely conceal the wiring path when aesthetics are a consideration. Flexible cable loops are used when a seamless appearance isn't necessary and must only be exposed on the secure side of the door.

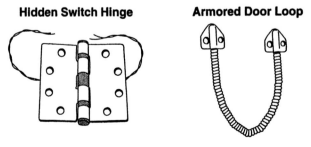

Electric Dead Bolt Lock

The electric dead bolt refers to the *bolt design* and is used as an alternative to a magnetic shear lock for doors that swing in two directions and double-doors. The electrically powered dead bolt is fitted into either the jamb or the door itself and when activated, it protrudes (shown on previous page) or swings (below) into a mortised strike plate on the adjoining surface.

The dead bolt does not give way with a spring action. Once it is clicked in place, it stays in place until unlocked.

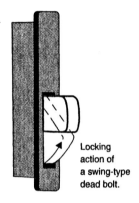

Locking
action of
a swing-type
dead bolt.

To increase holding strength, more than one set of electric dead bolts can be installed per door. Dual sets are common on large doors, as well as on both double-hung doors that swing away from each other from a center point. By installing electric dead bolts in the door header (top) and at the base, each door is secured and resistant to force.

Although electric dead bolts can be set in fail-safe or fail-secure modes, the majority of building and safety codes prohibit them for egress path use in high-rise buildings. Manufacturers have developed standard-compliant locks, but they are not in common use for these applications.

Fire Exits and ADA Rules

The rules surrounding fire exits sometimes conflict with the purpose of EAC.

No one wants to be trapped inside of a building during an emergency. This means that specific exits — doors leading to and from stairwells, between firewalls (and adjoining buildings), and directly outside — must be:

- Easy to see

- Easy to open in one simple motion

- Designed with minimal hardware (that is, a smooth surface with only one opening device)

- Latched in a fail-safe mode (that is, "not locked" from the inside)

- Closed immediately when released (have automatic door closers)

- Constructed out of fire-rated materials

Here is how fire codes effect EAC: In this simple example, the door is secured by a magnetic lock that can sense when the door is closed. To enter, a card is swiped through a card reader which sends the information found on the card to a control panel. If the card is valid, the control panel sends the instructions to unlatch the lock.

After the door is opened, the "door closed" sensor tells the control panel whether or not the door returned to the closed position. If the door does not close within a predetermined amount of time, the control panel triggers an alarm.

- Exterior View -   *- Interior View -*

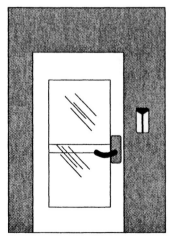
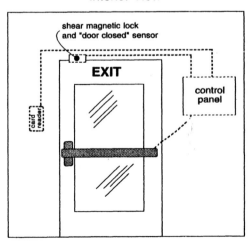

Whether or not the door closes as scheduled, the EAC database saves the passcode user's name as well as date and time of his or her access. This creates an important trail of information!

Exiting, however, creates a different set of circumstances.

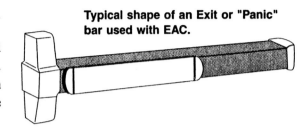

Typical shape of an Exit or "Panic" bar used with EAC.

The Exit Bar in this example sends a signal to the control panel. The control panel then releases the magnetic lock. Unfortunately, this action leaves *no record* of the person who pushed the door open, because exiting bypasses the EAC recording system.

The Americans with Disability Act (ADA) imposes additional restrictions on door design, lighting, and usage. ADA requires that:

- Blind and sight impaired people must be able to touch specific types of door hardware and understand what to do next.

- Hearing and sight impaired people must be able to easily see exits. Consequently, there are rules regarding the size and color of exit signage, including the use of strobe lights.

- Physically weak people as well as those confined to wheelchairs must be able to push a locked door open with little or no trouble, eliminating knobs and multiple latches.

- Wheelchair confined people require doorways with clearings of at least 32 inches, which is room for a wheelchair to pass.

Exiting, obviously, opens previously secured passageways.

To alert guards that someone is leaving, an egress button is sometimes found on the opposite side of a door protected by EAC. *When pushed, this button disarms an alarm and tells the control panel that door usage is in compliance with the system.* Egress buttons, unlike card readers, are subject to fire code regulations that forbid them to control locks. Egress buttons, therefore, can be bypassed without hampering travel, although doing so will trigger an alarm.

A "delayed egress" device on a fire exit door, however, postpones unlocking for up to 15 seconds. Pressing this device sends an alarm to a guard station and informs the guard that an exit attempt is being made. At this point, the guard can see the exit event on CCTV, talk to the person leaving through an intercom, or simply run to the scene if neither of those devices are there.

Obviously, a 15-second delay in exit can be frightening in an emergency situation, especially if the person attempting egress does not know what is happening. Extreme care must go into designing this type of exit system, which includes posting bold warning signs. *A single-push bar egress system is required even when delayed action is used.*

It is very common to see fire code violations and when you do, it is our strong recommendation that you immediately report them to the fire department. The National Fire Protection Association (NFPA) code clearly states that only one action can be used to unlock a door with exit or fire exit hardware. Many companies, unfortunately, install additional locks. If the lights fail during an emergency, the extra burden of finding those locks could cause confusion, panic, and death.

Double-exit doors, where one door must be opened before the other is released, are forbidden. In the case where the doors have an overlapping astragal (center strip), which normally requires one door to open before the other, hardware must be installed that allows either door to open quickly. Locking arrangements on double-exit doors are tricky and mistakes are often made during installation. Always check to see that each door can be opened quickly, regardless of the other's position. If one doesn't open, the setup is in violation of fire code.

Heavy double exit doors are commonly seen in shipping and receiving areas. The temptation is to install additional handles to better distribute the weight of the door in order to make opening easier. This solution, however, would be in violation of fire codes. In the event of an emergency, it might not be obvious which handle is associated with the latch, which could, in turn, cause confusion and panic.

Stairwells pose additional security concerns. Fire codes require that people in stairwells be able to exit freely at any floor. Unfortunately, in some high rise buildings, these exits open into unrelated businesses. The temptation is to bar the exits to stairwell doors so that uninvited guests don't get in, which, of course, is in violation of fire code.

As you can see from this brief overview, building codes, fire regulations, and ADA requirements are detailed and complex.

Internal Gates and Turnstiles

No matter how complex an EAC system is, "piggybacking" or "tailgating" may be a problem. This happens when one authorized person uses his EAC credential to open a door and then bypasses further access control procedures by admitting additional people before the door is shut.

Turnstile with Card Reader

Gates and turnstiles decrease casual piggybacking by forcing people to enter one-by-one. This enforcement is greater when a gate or turnstile is positioned at the beginning of a narrow lane (also called an *alley*) which is designed to hold only one person at a time. Sensors strategically embedded in lane walls are programmed with a *delay feature* based on the time it takes an average person to travel through the enclosed area. If the lane is violated by nonstandard use, such as someone stopping in the lane, or more than one person at a time, an alarm is triggered.

Example of Lanes

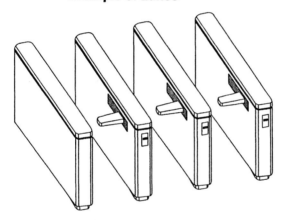

Unfortunately, poorly planned gates, turnstiles, and lanes can slow traffic.

To increase processing, multiple units are often installed to move high traffic and, to save energy, are selectively closed during low volume periods. Access to the Metro (subway) system in Washington DC, for example, uses an efficient multi-lane system. It requires that each passenger insert a magnetic striped fare card at a gate positioned at the beginning of a lane. Once the card is validated, the gate retracts and the passenger walks along the lane's area. Upon exiting the lane, the passenger retrieves his or her fare card and then, the process repeats itself for the next person in line. The whole thing takes place within seconds.

Mantraps *(secure vestibules and turnstiles)*

Tight access control is obviously very desirable in high crime areas. Financial institutions, hit hard by increased robberies, are exploring ways to quickly screen visitors. Many European and South American banks, for example, are using glassed-in mantraps, called "double vestibule (hall) portals," as seen in the illustration. These are used to unobtrusively examine visitors prior to admission, keep nonconforming people out, and make sure that two people do not enter at one time (piggybacking) as described in the following procedure:

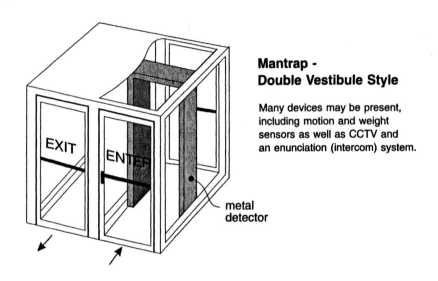

**Mantrap -
Double Vestibule Style**

Many devices may be present, including motion and weight sensors as well as CCTV and an enunciation (intercom) system.

metal detector

Entering a building:

> With the exterior (outside) door unlocked and the interior door locked, sensors and a metal detector determine whether *one person* is present in the "enter hall" and is *free of weapons*. When access is granted, the exterior door locks and simultaneously, the interior door unlocks. This allows the occupant to enter into the building while at the same time preventing piggybacking. The system resets itself when the interior door is shut, allowing the next person to enter from the outside.

Leaving a building:

> The person exits through the "exit hall," which reverses the door locking and unlocking process as reported above; however, does not include a weapon detection sensor.

David C. Smart, CPP, in the October 1994 issue of *Security Technology & Design*, reports that The Marshall Field Garden Apartment Homes, a low-income housing project in a high crime area in Chicago, has further refined this system. In order to pass through the mantrap, authorized people must also use a PIN and biometric handprint scan that is associated with their apartment number. Temporary visitors use a time-limited version of this same identification process.

According to Smart, strict access control in this housing project has greatly reduced the number of people freely roaming the halls and has increased the tenants' feelings of security. In one case where a rape did occur, EAC records were checked and a visitor was quickly identified, found, and hauled off to jail.

One concern is that biometric scanning might interfere with American civil liberties. Smart indicates that the palm prints used by this system are not used in the judicial system. Care must be taken when installing a mantrap, however, to make sure that it meets all fire and safety regulations and it does not interfere with the public's civil rights.

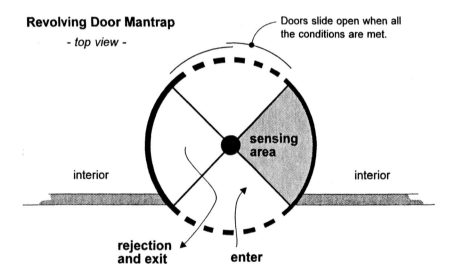

Revolving Door Mantrap - top view -

Revolving Doors: Revolving doors can also be used to reduce piggybacking and pose as mantraps. Used with or without an EAC passcode, one section of the area can be set up to sense for metal detection and other conditions. If all conditions are met, a person can pass through the system. If conditions aren't met, the interior doors remain locked and the person is directed back to the outside. As revolving doors are confining and have been known to cause feelings of panic, extreme care must be taken when using this type of system to meet all fire and ADA regulations.

Chapter Review - Barriers

To be regarded as a barrier, it must:

1. Define clear boundary markings of the area to be protected.

2. Delay unwanted traffic, but not necessarily stop it from happening.

3. Direct traffic to the proper entrances.

To be highly secure, barriers require . . .

continual surveillance. This is because most barriers can be defeated by ground weapons, natural growth, aviation, and the weather.

Vehicle gates controlled by EAC can slow down traffic. To avoid this . . .

install *continuous-motion vehicular access control.* This uses proximity or bar code technology that can check traffic at speeds of up to 30 MPH from a distance of 30 feet within a split second.

Four components every door controlled by EAC must have:

1. A door closing mechanism.

2. An electronically or magnetically activated lock.

3. Sensors (switches) that determine whether the door is open or properly closed.

4. Computerized control.

Four EAC installation considerations are:

1. Availability of sufficient electrical power for each door.

2. Sufficient wiring ducts and conduit in the walls and ceilings in compliance with local regulations.

3. Option for wireless EAC when the cost of installing wiring is prohibitive.

4. Sufficient cavities in the walls or ceilings to house EAC control panels.

Spring-activated tension in door closers must:

1. Have sufficient force to properly close the door after each use.

2. Make door opening very easy.

3. Be gentle on the door itself and not cause warping.

Two main categories of door closers:
1. Concealed (decorative and uncommon).

2. Surface mounted (very common).

Three types of surface-mounted door closers:
1. Regular-arm mounted.

2. Top-jamb mounted (the reverse of the regular-arm style).

3. Parallel mounted (least popular of the three, normally used when the arm must be exposed to the weather-side of the door).

Strong electronic and electromagnetic locks should resist:
1. Picking, which can manipulate the parts.

2. Drilling, which can destroy the lock.

3. Electronic or magnetic trickery, including the use of unauthorized passcodes or PINs.

Two very important locking conditions that can impact fire and other safety rules:
1. *Fail-safe condition:* The lock is **unlocked** when the power is off.

2. *Fail-secure condition:* The lock **remains locked** when the power is off.

Four common types of locks used in an EAC system:
1. *Magnetic:* Extremely popular due to low maintenance and high reliability. Comes in two types: shear (hidden) and direct hold.

2. *Electric Strike:* The most popular EAC lock on the market today. Can be used with either latch bolts or dead bolts.

3. *Electric Lockset:* Similar to standard mechanical locksets, except the key action is replaced by electronics. Comes in cylindrical and mortise styles.

4. *Electric Dead Bolt:* Usually used as an alternative to the magnetic shear lock for doors that swing in two directions and double-doors.

Six commonly known rules demand that fire exits be:

1. Easy to see.

2. Easy to open by a single action.

3. Designed with a smooth surface that has only one opening device.

4. Latched in a *fail-safe* mode (that is, "not locked" from the inside).

5. Closed immediately when released.

6. Constructed out of fire-rated materials.

Four commonly known rules imposed by the Americans with Disability Act (ADA):

1. Exits should be easy to see and attract attention, such as through strobe-lit exit signs.

2. Doors should be easy to push open to accommodate people in wheelchairs as well as people who are physically weak.

3. Doors can have only one easy-to-use unlocking device.

4. Doorways must provide a clear opening of at least 32 inches, which is room for a wheelchair to pass.

When pushed, egress buttons, sometimes found on the opposite side of EAC controlled doors . . .

> signal the control panel that opening the door is in compliance with the system and cancels any alarms that might be associated with the unauthorized use of the door.

Delayed egress on a fire door requires . . .

> that the door displays a warning sign, it opens within 15 seconds, and that the delayed egress device itself is controlled by a single, easy-to-use unlocking device.

Lanes with gates and turnstiles provide . . .

> protection against piggybacking.

Two primary considerations when installing a mantrap are that:

1. It meets all fire and safety regulations.

2. It must not interfere with American civil liberties.

QUESTIONS

1. What are four components of every door controlled by EAC?

2. What are the two main categories of door closers?

3. What is the difference between a fail-safe and fail-secure condition?

4. What are the four types of locks used in an EAC system?

5. What are six commonly known rules that govern fire exits?

Chapter 5

Sensors

Information Reporting Devices

This chapter provides an overview of sensor technology commonly used in electronic access control and in security systems.

For your reference, the last section of this chapter contains a glossary of terms that describe this technology.

Sensors Provide Input for Electronic Decisions

The "control" in electronic access control (EAC) is accomplished by the relationship between three types of devices, which are:

Detection Devices: These devices detect and report changes in one or more conditions. They are regarded as *inputs* because they report "into" a management device.

A Management Device: This is a specialized computer that receives information from detection devices, compares that information against programmed information, decides what to do, then issues instructions to action devices.

Action Devices: These devices carry out the instructions provided by the management device. They are regarded as *outputs* because the management device sends information "out to" them.

In a large EAC system, detection and action are managed electronically through control panels. These panels:

- Accept input from many detection devices.

- Issue instructions to many action devices (outputs).

- Communicate with other control panels and computers throughout the system.

Among the inputs we've studied so far in this book are authenticators and keypads. Equally important, but often invisible to us, are *sensors* and *detection devices*. In secret, these devices provide information about conditions upon which electronic decisions are made.

Historically, old-time intrusion detection systems were mechanical. Doors were rigged with all kinds of levers and pulleys that would trigger bells and/or start a chain of events:

> *Example:* To deter intrusion in castles, stones would drop through shoots, spears would fly out of walls and trap doors would open up, tumbling unauthorized people into pits full of snakes (or bodies).

Fortunately, electronically-powered intrusion detection systems (also called *burglary detection systems*) alert guards to a wide variety of issues without destroying an unaware visitor. Today, EAC uses the information reported by sensors to make informed decisions.

> *Example:* When an EAC controller receives a signal from a door contact sensor, it knows that a door was opened. If that signal was received after a proper signal from an authenticator, the controller regards the situation as being OK. If the contact sensor does not close within a specified time, the controller signals an alarm, which can include ringing bells, flashing lights and warnings seen on central station monitors.

Here are a few examples of how controllers use sensors to monitor situations:

- **Elevator Door:** An EAC authenticator determines who can select a given floor. Contact switches determine whether an elevator door is completely opened or closed. While closing, one or more photo electronic sensors determine whether people and/or objects are between the sliding door and the frame. Finally, pressure-sensitive sensors determine whether someone or something is attempting to hold the door open.

- **Exterior Door in a Chemical Plant:** An EAC authenticator determines who can access this door. One or more contact switches in a door frame determine whether the door is opened or closed. A contact switch that is part of the latch determines whether the latch is fully extended.

  Chemical and/or oxygen sensors inside the plant sense whether a chemical spill has taken place. If one has, the door will not unlock, even for an authorized person trying to enter.

  Photoelectric sensors around the door determine when people or objects are in the area. These sensors start a video recorder and/or turn lights on so the camera and visitor can see better.

The way sensors are used in a facility depend on overall security requirements and management needs. At minimum, a door controlled by an EAC system requires at least one door contact sensor (input), an authentication device (input), and an electronically activated lock (output).

Sensor Standards

At issue when selecting any sensors for a system is whether they work reliably. If they do not, they will cause a chain of events that report false alarms.

Just because a sensor never reports a false alarm, however, does not mean it is reliable. Failure to report problems only provides a false sense of security. In some ways, this situation is worse than one in which false alarms are constantly ringing because you are kept uninformed.

Troubleshooting Tips:
Establish procedures for regularly testing all sensors (also called *"alarm points"*) in your system.

Keep in touch with your sensors' manufacturers to learn about any changes in device design.

Check with other users of the same technology to learn from their experiences.

Read related magazines to keep on top of the subject of intrusion control.

New sensor technology is introduced daily, consequently, not all sensors have been thoroughly tested before you buy. Although there are no universal standards, efforts have been made to rate and/or certify equipment. To check on sensor specifications and standards, consult the following organizations:

The **U.S. General Services Administration** publishes specifications for alarm systems based on governmental procurement needs. These include troubleshooting information designed to help you test the system. At this writing, W-A-450C was available.

Underwriters Laboratories (UL) in Northbrook, IL prepares manufacturing standards and then certifies whether devices submitted for examination meet these standards. They publish a catalog that lists available reports on burglar alarm units, central-station units, intrusion detection units, etc.

The **American Society for Testing and Materials (ASTM)** in Philadelphia provides information on building security standards, which includes considerations for intrusion detection systems.

Sandia Laboratories, Albuquerque, NM also publishes various handbooks on detection and sensor systems. While these are not standards, they do provide solid overviews.

Sensor Categories

Sensors are either active or passive.

> **Active sensors** introduce energy into an area which is interpreted by a receiver. When the receiver senses a change in energy, it registers an alarm. Break-beam sensors are a good example of this type. Here, a transmitter focuses light, which is energy, into a receiver. When the receiver notes a change in light, such as when someone passes by and "breaks the beam," it triggers an event.

> **Passive sensors** measure changes in an environment over time. A good example of this type of sensor is a thermostat which, when the temperature drops, triggers a furnace. Likewise, passive sensors can detect noise and vibration levels within an area and indicate when those levels are outside a given range.

Sensing devices commonly used for EAC generally fall into the following categories. (Check the sensor glossary at the end of this chapter for specific types of sensors.):

- *Mechanical* sensors have simple levers or rods that, when pulled or pushed, move a switch that reports an event.

  > *Example:* An *egress* (exit) button used on the secure side of a door controlled by an authenticator is a mechanical switch that, when pushed, creates a electrical circuit that tells a control panel that opening the door is legal. If the door is opened, but the button isn't pushed, other sensors in the system announce an alarm.

- *Electromechanical* sensors depend on a specific flow of current to activate a mechanical device.

- *Capacitance* sensors generate an evenly charged electrical field between two antennas. When the energy level in that field changes due to an intrusion in the area, an alarm is triggered.

- *Vibration* sensors measure subtle environmental motion, which, when motion reaches a predefined level, register an alarm.

- *Audio* sensors are similar to vibration sensors, except they measure sound waves (audible, which we can hear and ultrasonic and microwave, which we can't).

- *Light* sensors measure the degree of light in a given area. Active light sensors are used in break-beam devices wherein the interruption of the light beam between a transmitter and receiver results in an alarm. Passive light sensors measure environmental light.

Additional sensor categories exist for environmental monitoring, such as for fire, flood, humidity, oxygen and chemical detection, to name a few. All these sensors can be tied into a controller of some type (including some EAC controllers) to automate a chain of events.

Sensor applications can be quite complex. In fact, many systems maintain redundancies, which means that one variety of sensor checks on another in order to double-check intrusion reports. With that in mind, the brief list that follows shows what types of sensors are commonly used within specific areas.

Yards - External Perimeter: Fence alarms (conductive wire sensors), photoelectric beams and microwaves.

Building Perimeter: Exterior door contacts and overhead door contacts (contact switches) and glass break detectors.

Interior Detection: Passive infrared, microwave, dual motion technology, photoelectric beams, interior door contacts, mantrap components, and glass break sensors.

The section that follows is a Sensor Glossary, which provides an overview of the types of sensors used in EAC and security systems.

Check the Appendix of this book for references to where you can get more information on this subject.

Sensor Technology Glossary

This glossary is meant to provide an overview of sensing technology terminology commonly related to EAC and intrusion detection systems.

Within a sensor type, there can be many variations. Check with sensor manufacturers for details.

There are many sensor systems not mentioned in this glossary that are commonly used in industry. We recommend that you become aware of them. The more you know, the more resourceful you'll become when designing a system.

Active System
(*Example:* Capacitance Detector) The word "active" refers to a sensing system that introduces energy through a transmitter into an area for interpretation by a receiver. The receiver is set up to expect a specific energy level. Any changes to that energy level indicate a change in the environment caused by an invasion of some type. The opposite is a *passive system*, which simply reads the environment "as is" and makes decisions based on a range of outcomes, such as increased noise or impulse.

Audio Sensors
These sensors are similar to ultrasonic and microwave sensors, except that the receiver bases its judgment on sounds that can be heard by the human ear, rather than a high frequency pitch. Reception sensitivity can be set to detect explosions, gun shots and even human conversation. (See *Ultrasonic and Microwave Sensors.*)

Audio sensors are usually used in connection with intercom systems and can amplify low noise, such as whispering, for transmission to remote guard stations.

Two types of audio sensors exist. The first is sensitive to sound at any frequency within a range. The second is sensitive to sound at a specific frequency.

In high background noise applications where vibration sensors are used, *discriminator sensors* are also installed as a redundant backup. These devices sense common noise and cancel the effect of these vibrations, reducing false alarms based on common occurrences. (See *Vibration Sensors.*)

Balanced Magnetic Switches
See *Intrusion Switches.*

Break-Beam Sensors

See *Light Sensors*.

Capacitance Detectors

This type of sensor is used to monitor large areas by maintaining a consistent energy level, called an *electrical field*, between two electrically charged antennas. The air in the electrical field becomes a *dielectric space*, meaning that it has a constant, predefined energy level. When the energy level changes due to an intrusion, an alarm is sounded.

While capacitance sensors are not affected by noise or vibration, they are very sensitive to atmospheric changes and consequently, are most commonly used indoors.

This type of sensor is relatively easy to set up by taping antennas of copper tubing or wiring to windows, walls, door frames, etc. As long as the energy within the room between these antennas remains constant, the area is secure. Intruders, however, absorb part of the radiated energy, causing a difference in *capacitance* (electrical charge), which, in turn, triggers an alarm.

Conductive Wire Sensors and Fiber Optics

Metallic tape, once commonly seen on glass doors and large windows, carries a *current* (indicating conduction) that completes a circuit. If the tape is broken, an alarm results. Unfortunately, although it is easy to apply tape to glass surfaces, it is also very easy to scratch through the tape, causing a complete split that breaks the electronic circuit. This renders the system useless and in need of continual repairs.

Another type of conductive wire sensor is a fine, hard-drawn copper wire that is woven into screens, grids and other lacings (such as used in fencing) and mats. Changes in tension on the wire, such as caused by someone pressing on a surface, changes current flow, triggering an alarm.

In newer systems, fiber optic filaments are used, with light transmission replacing electrical conduction. The principle, however, is the same as with conductive wire. Fiber optics eliminates corrosion problems common with metallic materials and is especially useful in outside applications.

Discriminator Sensors

See *Vibration Sensors*. Also see *Audio Sensors*.

Dual Motion Detectors

This refers to a redundant system in which one type of motion detection system backs up another. Either system can be used by itself, but when used together, they provide a broader, more complex range of coverage. Generally a dual motion detection system combines passive infrared (PIR) and microwave

(MW) motion detectors, or PIR and ultrasonic (US) motion detectors. A discussion of these technologies is seen under *Ultrasonic and Microwave Sensors* and *Light Sensors.*

Electrical Intrusion Switches
See *Intrusion Switches.*

Fiber Optic Sensors
See *Conductive Wire Sensors and Fiber Optics.*

Fire and Environmental Sensors
Environmental sensors can be tied into electronic access control systems, however, local fire, building and police authorities determine usage and insurance companies may insist on additional requirements.

Among common sensors used for environmental purposes are smoke detectors, heat/temperature sensors, chemical spill detectors, water flow monitors and moisture detectors.

Flexible Cable Sensors
See *Conductive Wire Sensors and Fiber Optics.*

Foil
See *Conductive Wire Sensors and Fiber Optics.*

Glass Break Sensors
The original glass break sensors used conductive foil taped along the side areas of the glass. As this type of sensor was easy to scratch, it caused many false alarms and is now considered obsolete. (See *Conductive Wire Sensors and Fiber Optics.*)

Today, a popular sensor used to detect glass breaking is a small capsule containing liquid Mercury (a conductive metal), which is glued to the glass. Once the glass breaks, vibrations and/or dropping causes the Mercury to flow across the capsule, closing a circuit, which, in turn, sends an alarm.

Sensors that fall in the sound and vibration categories are also used for glass. These sensors can be tuned to audible or vibratory frequencies that match the frequencies of glass breaking. (See *Sound Sensors.* Also see *Vibration Sensors.*)

Infrared
This refers to the part of radiation within the full radiation spectrum that falls below visible light. It can't be seen by the human eye, but it can be felt as heat. Sensors that detect infrared heat detect the presence of warmth, such as that radiated by a human being or animal.

Infrared Sensors

See *Light Sensors.*

Infrasonic Sensors

These are sound sensors that detect sound below that detectable by ear. They can, for example, "hear" the sound of air moving into a room when a door is opened. They are not widely used today, however, because of false alarms.

Intrusion Switches

This type of sensor can be mechanical (similar in concept to a rocker switch used to turn on lights) or electrical. It has two parts, typically one that moves or changes state and one that interprets the change.

Electrical intrusion switches are commonly installed on the secure-side of windows, doors and other openings. Under normal conditions, both parts of the sensor touch, creating a circuit through which current flows. When separated, such as happens when a window is illegally opened, the flow of current is broken, signaling an alarm.

A variation of this type of switch is the *magnetic switch*. This switch, which is an electrified plate, is mounted on a fixed frame, while a non-electrified metal plate is mounted on a moveable object, such as a door or window. When the two plates contact, a stable magnetic field registers. When separated, the magnetic field is disturbed, causing an alarm.

Laser Sensors

See *Light Sensors.*

Light Sensors

Light sensors respond to changes in light level and are used in a wide variety of industrial applications in addition to EAC.

One type, called an *ambient light sensor*, measures daylight. When this sensor detects that daylight is dimming, a controller responds by turning on lamps. If lamps are not controlled by timing devices, then ambient light sensors are most likely being used. You usually can tell when one is present on a light pole by seeing a small dome on the very top of the lamp fixture.

In security applications, photoelectric sensors are commonly used. Known as *break-beam sensors*, they consist of two parts: A transmitter and a receiver.

The transmitter beams a tightly focused beam of light at the receiver. When the beam is broken by someone passing between the transmitter and receiver, the receiver notes the change, then triggers an event. These events can include sounding an intrusion alarm, triggering video recording, or opening a gate when someone or thing approaches, then shutting the gate as soon as it's clear.

Photoelectric sensors use specific types of light. These include light generated from specially designed incandescent bulbs, infrared light or laser light. No matter what source is used, the light transmitter is adjusted to tightly focus the light beam on the receiver.

Depending on the type of sensor, intruders can defeat break-beam sensors by fixing a flash light on the receiver. To solve this problem, light transmission is commonly *modulated* (pulsed) in a way that cannot be duplicated by a constant light beam from a flashlight.

Technology is improving the way photoelectric sensors work. Intruders can be tracked, for example, by zigzagged light paths rigged through a system of mirrors. In addition, laser beam sensors are increasingly replacing infrared due to greater beam strength and focusing capability.

Magnetic Switches
See *Intrusion Switches*.

Mats (pressure)
See *Pressure Mat Sensors*.

Metallic Tape
See *Conductive Wire Sensors and Fiber Optics*.

Microwave Sensors
See *Ultrasonic and Microwave Sensors*.

Motion Detectors
See *Ultrasonic and Microwave Sensors*. Also see *Video Motion Detectors* and *Dual Motion Detectors*.

Passive System
(*Example:* Passive Infrared.) The word "passive" refers to a sensing system that measures changes in the environment over time. This could include changes in infrared light, temperature and humidity normally found within an environment. The opposite is an *active system*, which actively introduces energy into the environment through a transmitter for interpretation by a receiver.

Photoelectric Sensors
The "photo" in the word "photoelectric" refers to light. See *Light Sensors*.

Pressure Mat Sensors
This type of sensor mat is used in man traps, entrances and exterior yards. They trigger an alarm when a specific weight (from 5 to 20 pounds per square foot) presses on the surface. Fiber-optic mats are preferred for outdoor or moist applications. Also see *Conductive Wire Sensors and Fiber Optics*.

Proximity / Capacitance Detectors
See *Capacitance Detectors*.

Sonic Sensors
See *Ultrasonic and Microwave Sensors, Audio Sensors,* and *Infrasonic Sensors*.

Timed Applications
When a control panel receives an alarm from a sensor, it may time how long the sensor remains in an alarm state. If a sensor returns to normal within a predetermined time span, no alarm is sounded.

Timing is used to measure the travel time a person or vehicle requires when passing through an access point. If the time set is too short, false alarms occur. If the time is set too long, it does not become obvious when a door or gate is improperly held opened.

Ultrasonic (US) and Microwave (MW) Sensors
These sensors measure ultrasonic sound and microwave energy. Ultrasonic sound is lower on the frequency scale, but above our threshold of hearing, while microwave energy is higher than ultrasonic and is regarded as electromagnetic energy.

Ultrasonic and microwave sensors work on a similar principle. They broadcast sound at a specific frequency, which is picked up by a preset receiver. As the broadcast is spread over a specific area, anything moving within that area disturbs the frequency pattern. If the receiver picks up a frequency that is different from what it expects, an alarm is sounded.

The broadcast frequency can be adjusted to allow for probable disturbances, such as small animals or birds. It can also be set to distinguish the stride rate of a moving person, sounding an alarm within four consecutive steps.

Ultrasonic sensors are normally used to monitor interior spaces because their frequencies are easily disturbed by the environment. Their frequencies must be adjusted with regard to the presence or absence of furniture, as materials absorb sound and alter frequency wave lengths. Generally, interior ultrasonic sensors are stable. Air currents caused by air conditioners, however, can set off false alarms.

Microwave sensors are better suited for outdoor use and are employed in sensing the sky at airports as well as sensing land use around remote prisons and military bases.

Vibration Sensors
In stable environments, this type of sensor samples vibration rates. Should the normal atmospheric vibration rate change, such as caused by cutting, chisel-

ing, or ripping, an alarm is sounded. This type of sensor is well suited for installation on masonry walls because masonry is naturally low in vibratory properties, thus reducing the probability of false alarms.

In applications with higher vibratory background noise, *discriminator sensors* are also installed. These devices sense common noise and cancel the effect of these vibrations, reducing false alarms based on common occurrences. (See *Audio Sensors.*)

Video Motion Detectors (VMD)

This sensor electronically analyzes CCTV camera images. It detects changes that are judged large enough to warrant an alarm. In a digital system, this detector notes changes in light levels from one set of *digital pixels* (square units) in a TV frame to similarly placed pixels in the next. An intruder casting a shadow over the area, wearing clothing with a different light refraction than the background materials and/or illuminating the area with a flashlight would cause this detector to sound an alarm.

In an analog CCTV system, the detector compares large areas in one frame to the same areas in the next. Analog comparison is more susceptible to false alarms caused by lighting changes and camera vibration than in a digital system, however, and is not recommended for outdoor applications.

Chapter Review - Sensors

The three types of devices required for electronic decisions are:

1. Detection devices, such as authenticators, keypads and sensors.

2. A management device, which is a specialized computer.

3. Action devices, such as locks, lights, buzzers, etc.

Sensors are regarded as *input devices* because they . . .

> report *into* a management device.

Locks, lights and buzzers are regarded as *output devices* because the . . .

> management device *outputs* instructions to them.

Control panels electronically manage EAC situations by:

1. Accepting input from many detection devices.

2. Issuing instructions to many action devices.

3. Communicating with other control panels and computers throughout the system.

Unlike authenticators, such as card readers, sensors are special in security settings because they are usually . . .

> hidden and report their findings secretly.

Sensors that do not work reliably cause . . .

> false alarms, which waste resources and/or no alarm during a problem, which keeps you uninformed.

Sensor technology is either:

1. Active, meaning that the sensor introduces energy into an area.

2. Passive, meaning that the sensor measures elements in the environment over time.

Chapter 5

Six general categories of sensing devices commonly used for EAC and security are:

1. Mechanical sensors.

2. Electromechanical sensors.

3. Capacitance sensors.

4. Vibration sensors.

5. Audio sensors.

6. Light sensors.

Mechanical sensors . . .

> have simple levers or rods that, when pulled or pushed, move a switch that reports an event.

Electromechanical sensors . . .

> depend on a specific flow of current to activate a mechanical device.

Capacitance sensors . . .

> measure the constant electrically charged field between two antennas and report a problem when that field changes.

Vibration sensors . . .

> measure subtle environmental motion.

Audio sensors . . .

> are similar to vibration sensors, except that they measure sound waves. These waves might be audible, which we can hear, or ultrasonic and microwave, which we cannot.

Light sensors . . .

> measure the degree of light in a given area. These sensors can be active or passive.

Three common area groupings in which sensors are used are:

1. Yards - external perimeter of a facility.

2. Building perimeters - exterior surface of a building.

3. Interiors - interior rooms of a building.

QUESTIONS

1. A control panel is a _____ device.

2. Inputs and outputs control a chain of events. What exactly is an "input" and an "output"?

3. What are the two main problems that occur when a sensor fails to work correctly?

4. What is the difference between an *active* and a *passive* sensor?

5. What is the difference between a mechanical sensor and an electromechanical sensor?

Chapter 6

Computers

Software, Hardware and Intelligent Networks

Many of the concepts in this chapter refer to computers that run on Intel-type microprocessors and are known as "personal computers" or PCs.

These concepts are general enough, however, to give you an overview of computer hardware principles, no matter what system you have.

Why You Should Understand Computers

Many devices used in electronic access control (EAC) systems are controlled by microprocessor chips.

The most useful microprocessor-driven devices allow us to customize their behavior through software instructions. They can also communicate with other devices and calculate a wide range of information. We commonly call these devices *computers*. The three main types of computers used in a large EAC system are the:

Supervisory Computer   The supervisory computer is a personal computer (PC), with monitor and keyboard, used to manage an EAC network. It issues information and instructions to other computers on the network, receives reports from those devices and stores information about ongoing events.

Controller   (or *Control Panel*) The controller provides a direct link to electronic authenticators, locks, sensors, gates, etc., installed at the site. When connections permit, it can communicate with the supervisory computer, but does not normally have a dedicated monitor or keyboard.

Switcher   The switcher controls closed circuit TV cameras and video taping activities and is commonly linked to sensing devices. Like a controller, it can communicate with the supervisory computer, but does not normally have a dedicated monitor or keyboard.

Chapter 6

Few security professionals, unfortunately, have a formal education in computer technology, even though they've all used software. This forces them to rely on advice about computer hardware from non-security professionals; advice that may be at odds with actual needs.

This chapter, then, is designed to provide a quick, general education in computer technology as it relates to EAC. Later chapters look specifically at control panels and the devices they service.

Computer Study Tips:

- While we urge you to read security-related publications, we also recommend that you regularly skim *Byte* and/or *PC Magazine*. Both magazines do an excellent job reporting hardware developments; developments which eventually migrate to security applications.

- Periodically thumb through computer supply catalogs just to see what's up.

- Learning about computer hardware, unfortunately, means that you have to understand cryptic acronyms, such as DRAM, BIOS, etc. Make life easy on yourself by writing down technical acronyms and their meanings as you become aware of them. This saves you from being overwhelmed by technical details when making critical decisions in the future.

Identifying Your Computer Hardware

In an emergency, you must be able to accurately describe your computer, even if you don't understand all its parts. The descriptions you need are reported by diagnostic software found on your system. Print those descriptions before you have problems, not after, when this information might be stuck in a system that does not work.

Most computer operating systems come with some form of diagnostic reporting. In Windows 95, for example, these reports can be found by clicking the icon named "My Computer," then "Control Panel" and finally, "System."

Computers that run on DOS version 6.0 or higher (with or without Windows 3.1 or Windows for Workgroups) provide diagnostic software called *MSD* (Microsoft Diagnostics).

> **TO FIND OUT WHICH DOS VERSION YOU HAVE:**
>
> At a DOS prompt, type the word VER, then press ENTER to see the results.

Run MSD or other diagnostic software to answer the questions posed on the following chart and keep a copy of your answers by your computer. The rest of this chapter will help you understand the information requested.

> **TO RUN MSD:**
>
> At a DOS prompt, type MSD, then press ENTER.
>
> > If you are running Windows or Windows for Workgroups, go to Program Manager, select File, then Run. Type MSD, press ENTER, read the message, then press ENTER again.

Know Your Computer	
Items	**Your System**
Computer name: *This is often the name of the microprocessor itself, and not necessarily the brand name of the PC.*	
BIOS manufacturer:	
BIOS version, if any:	
BIOS category:	
Processor type:	
Bus type:	
Amount of extended memory (RAM):	
Operating system (OS) name and version:	
Microsoft Windows version, if present:	
Video type:	
Mouse type: (serial or parallel)	
Drive names, types, and sizes:	
Stacked (compressed) drives, if present:	
Sound card, if present:	
Com 1, Com 2, Com 3, etc. are used for: *Com 1 and 2 are typically used for a mouse and a modem. Com (communications) ports are also called "serial ports."*	
LPT 1, LPT 2, LPT 3, etc. are used for: *LPT refers to a parallel port commonly used for printers. The initials "LPT" stand for "line printer terminal."*	
Video capture board, if present:	
Other add-on components:	

Identifying Your Hook-Ups and Switches

Label your hardware, including cabling, connection points, buttons, switches, etc., so that people know what they are used for.

Examples of Common Problems Solved by Labels:

One major computer manufacturer placed its On/Off button very close to the floppy disk eject button. Without a label, users accidentally turn the computer off in the middle of an operation when removing floppies.

Another major computer manufacturer created identical On/Off and Reset buttons. Users often press both, hoping the computer will figure out what they need. Unfortunately, quickly flicking the On/Off button can jolt the hard drive, causing it to crash. Labels stop this from happening.

Make sure that you label both ends of every cable — such as "to printer" and "to computer." Inexpensive label kits are available from stores like Radio Shack, or you can make them yourself.

Labeling helps when you have to move your computer. It is invaluable to other people who use your system, especially when you have to instruct them over the phone.

Draw maps identifying computer connection points. Place those maps in obvious places, such as taped to the side of your computer and at every connection hub, control panel, and switcher throughout your system.

Tip: Once you start adding additional cards to your computer, it becomes very difficult to tell what all the outlets are used for. Keep your connection maps up-to-date, minimizing the need to fumble during an emergency.

Last, clearly label the circuit breakers in electrical cabinets. Avoid the problem of having an under-trained staff member resetting all the circuit breakers in an area because he or she cannot identify the problem.

Redundancy - Taking Extra Special Care

Very commonly, the command post for all your security information is a single computer. No matter how many computers you might have in your system, ultimately, one computer oversees the rest.

Backing up data on a daily basis is, of course, a must, but what happens when your supervisory computer fails?

While you should contact a top security consultant to determine how you can introduce *redundancy* (duplicated processes and equipment) into your system, most PC-based systems are small enough to enable this simple solution:

- Buy a second computer that duplicates your supervisory computer in every way.

- Duplicate all your software and data on this computer. This can be done through a communications link or through a backup device.

- If a problem occurs with your supervisory computer, substitute your backup computer during the repair process.

Maintaining a distributed system, which we discuss later in this chapter and again in Chapter 8, *System Design*, keeps your doors locked properly even if there is a problem with one of the units, adding a level of redundancy to overall security.

Also make sure that your lines of communication are redundant. Having multiple ways of contacting local authorities, such as by modem, direct line and wireless channels, is an important plus.

Use battery backup devices (not just surge protectors) to keep your system running during a storm and protect your equipment against damage from electrical spikes, which can damage sensitive electronics.

Last, keep a supply of fresh, new cables available and label them as to their use. Intermittent system problems can be caused by a single broken wire at a connection point in what otherwise looks like a strong, thick cable. Breaks can happen unexpectedly, such as when something is accidentally shoved against a plug.

Don't jeopardize security in the evening or over the weekend by having to wait until the next day for computer supply stores to open. In addition, do not keep old used cable for replacements. The few dollars saved might cost you hundreds, if not thousands, in headaches.

Troubleshooting Your Computer

Complaining that your computer doesn't work is as descriptive as saying your house doesn't work. When the computer doesn't work, try to isolate and describe the source of your problems to the best of your ability before talking to a technician.

It is important that you be as knowledgeable as possible. Computer technology is rapidly changing and no one person knows it all. Consequently, describing problems clearly and listening to replies intelligently is absolutely critical.

The following troubleshooting tips might prove helpful, whether or not you completely understand your computer system.

- **DON'T UNDERSTAND THE TECHNICIAN'S INSTRUCTIONS:**
  Before talking with a technician, make sure you *clearly understand* the following concepts:
  > Go to DOS
  > Go to a particular drive, such as C: or D:
  > Go to a directory
  > Find a file
  > Edit a file
  > Close or save a file
  > Rename, copy, move, or delete a file

  If you are running Windows, you still need to know the above, plus:
  > The directory structure as shown in File Manager or Explorer
  > How to maximize and minimize screens
  > How to use Notepad or other Windows text editors
  > How to switch between programs

- **CAN'T INSTALL A NEW DEVICE:**
  Brand new devices and software are often developed around the latest version of an operating system. Even though an old Disk Operating System (DOS) and/or networking program may be adequate for your current setup, it might prevent you from successfully adding new devices.

  Keep your computer up-do-date and replace it when necessary. Check whether your system has enough electronic memory. Often, the minimum "required" memory is not enough.

- **INSTALLED A NEW DEVICE, BUT CAN'T MAKE IT WORK RIGHT:**
  Before installing a new device, check the version numbers of all the software on your system with the manufacturers to make sure you have the latest. This includes special software required to run your monitor, sound card, video capture card, etc., commonly called *"drivers."* As

software products are becoming increasingly interrelated, an old version of one might negatively effect a new version of another.

Always double-check the version number and issue date of the software packaged with the device *you just purchased.* Even though the product is new, its software may not be the latest.

To check version numbers, call the manufacturers involved, check vendor bulletin boards, on-line services and/or related Internet Web sites.

If you are working with a dealer or installer, insist on verification of the latest version numbers. Also request that the dealer completely check and update your current system before adding something new.

Remember: Software can be updated several times a year before a completely new release is sold. You must be aggressive to stay current. Few manufacturers send out notifications of incremental software upgrades. Even if they do, sometimes dealers don't automatically pass this news along to the end user.

Redundancy Tip: Backup your system before adding a new device or software. This is especially important for systems using graphically-oriented operating systems.

- **A SUCCESSFULLY INSTALLED DEVICE, SUCH AS A PRINTER, NO LONGER APPEARS TO WORK:**
  At start-up, your computer checks to see whether various devices are attached. If a device was *off* when you started your computer, the computer may not detect its presence. Prior to starting your computer, always make sure your other devices are *on*.

- **A DEVICE WORKS INTERMITTENTLY, OR NOT AT ALL:**
  Check whether your cables are good. Tight bending and angles greater than 45 degrees can easily snap internal wires. Although the cables seem sturdy, they contain delicate, slender wires. Snapping a single wire may not be enough to disable a device, but it can cause intermittent problems.

- **A DATABASE IS NO LONGER ACCESSIBLE AND IS DISPLAYING AN ERROR MESSAGE:**
  See the *Database* section of this chapter.

- **CAN'T GET ANSWERS TO SOFTWARE QUESTIONS:**
  Buy software that is well supported by the manufacturer and is well understood by area technicians and consultants. Check to see whether the software developer has an adequately staffed help-line and whether calls are returned within 24 hours.

If the manufacturer has a Web site, see if you can post questions through Email. Not all sites feature this service. Also see whether the manufacturer has a presence on an online service, such as CompuServe or America Online.

Tip: Busy help-lines can take a half-hour or more to personally answer, meaning you are put on hold. If you must get an answer, place the call on a speaker phone. In this way, you can do other things while waiting for the technician to answer, without having to hang onto the handset.

- **CAN'T FIND GOOD TECHNICAL SUPPORT:**
  Buy hardware from stores or dealers who have a proven ability to provide technical help and troubleshooting. Talk with the service manager to find their average turnaround times. Also check how long he or she has been on the job.

- **DISAPPOINTED BY BARGAIN COMPUTER OR OTHER PRODUCT:**
  Make sure that the "good price" you pay for computers, software, or other devices is not based on obsolete products. Always ask "is this product the most recent version" and, if the store can't verify it to your satisfaction, call the manufacturer or ask questions on online services.

- **HAVE SLUGGISH PERFORMANCE EVEN THOUGH YOU HAVE A FAST COMPUTER:**
  When buying a computer, purchase at least double (better, triple) the minimum suggested electronic memory (RAM). The lack of sufficient electronic memory can cause *extremely* sluggish performance, even with the most up-to-date microprocessor.

- **HARD DRIVE DISK ACCESS SEEMS UNUSUALLY SLOW:**
  The computer's operation can be impaired if the hard disk is more than 80% to 90% full. Always check the amount of free space on the hard disk. Copy unused files to another media, then delete them from the hard disk to free up space.

  Although it is very unlikely that a computer used for security purposes would also be used to surf the Web, make sure that the directory containing your Web contact history isn't jammed with unneeded files.

  Be aware of software you are using that generates "log" or history files saved to the hard drive. If left unattended, these files continue to grow in size. Also be aware that today's modem operating systems use part of the hard drive in order to function.

- **HAVE TROUBLE LEARNING NEW SOFTWARE:**
  When trying to remember software steps, write them out cookbook-style on simple index cards.

Technical Information

Understanding the components of computers greatly aids the understanding of computerized electronic access control networks.

Operating System and Other Software

The *microprocessor* that manages information flow in your computer or intelligent device is itself managed by instructions.

These instructions are either *hardwired*, meaning that they are physically embedded in your system, or are issued by *software*.

Software is composed of written words, just like the words you're reading in this book. The actual language of software, however, is like any foreign language. You need to learn it in order to use it, consequently, most people regard software language as *code*.

The information you see on your monitor is an interpretation of software code translated into a readable format. While you might be required to program your EAC system, most likely you'll do so by answering questions with everyday language. Once you enter your choices, the software translates this information into its own code.

Computers use many layers of software, most of which is not apparent. The very first piece of software your computer uses is called the *operating system*, or OS for short.

The acronym DOS refers to *disk operating system*, meaning that the computer supports disk drives. While this may seem obvious, computers were once controlled by tape drives, similar to those used by tape and video recorders. The designation DOS on "new" machines at that time told the public that they were state-of-the-art.

There are a number of operating systems available, among which are: MS-DOS or PC-DOS, OS/2, WARP, UNIX and high-end Microsoft Windows products.

Low-end Windows products, such as Windows 3.1 and Windows for Workgroups, require that MS- or PC-DOS be installed before they will run.

Essentially, MS-DOS and PC-DOS are the same operating system. MS refers to the software sold by Microsoft and PC to software sold by International Business Machines (IBM).

Database Software

Next to your computer's operating system, database software is the most common type of software used in EAC.

Database software is especially good at:

- Acquiring a history of events,

- Creating and maintaining records related to authentication,

- Holding specialized information related to the way locks, sensors and other devices work, and

- Providing data storage and report generation.

Unfortunately, power failures, static electricity and turning off the computer before properly closing software can cause serious trouble. Database software opens and closes many files. When some files are saved, but others are not due to premature shut down, the database becomes corrupt.

Avoiding Corrupt Databases:

Solution 1: Plug your computer into an *uninterruptable power supply*. This is an electronic device that places a battery between the wall current and your equipment. If a storm cuts power, your computer will continue to run flawlessly and your database will be safe.

Solution 2: Most database software lets you *repair* or *rebuild* your files if they are damaged. Take emergency precautions and understand how to rebuild files before this type of accident happens. Post these instructions on your computer so your operators recognize signs of the problem and can quickly correct the error without causing lost time and confusion.

Solution 3: Always make **daily backups** of database files. The best way to backup is to rotate a series of 7 to 14 backup tapes (or disks), rather than only one or two. Do this to protect against virus invasions which may be present on fresh backups.

Solution 4: File errors can also be caused by the failure of your computer's clock battery. Although your computer can run without a clock battery, once the computer is completely turned off, the current time and date will *not* be available when it is turned on again. In this case, manually reset your computer's date and time before starting the database software. Next, call a technician to change the battery because this requires special tools.

Microprocessor

The operating system, your software, plus your unique information are all manipulated by the computer's *central processing unit* (CPU), which is another name for "microprocessor."

Inside your microprocessor are tens of thousands of transistors.

A transistor is a microscopic electronic switch that has two potential states (positions). It is switched *on* when it is energized (passing on a jolt of electric current) and *off* when not. Transistors control many EAC devices, allowing these devices to be powerful but small.

How Many Transistors in a Microprocessor?

Microprocessor Name	Date Introduced	Transistors in the Microprocessor
XT (8086)	July 1978	29,000
AT (286)	February 1982	134,000
386	October 1985	275,000
486	April 1989	1,200,000
Pentium	March 1993	3,100,000
Pentium Pro	September 1995	5,500,000

NOTE: Pentium and Pentium Pro are trademarked names of the Intel Corporation.

The benefit to using transistor technology is that it provides an extremely fast rate of information exchange, while being packaged in a very small unit. Unfortunately, because transistors are 100% dependent on electrical power, changes in power supply cause problems:

- When the power shuts off, all switches return to the off-state, losing all information.

- When static electricity strikes, the switches retain their state, but do not have enough power to change. This causes the computer to appear frozen.

All information interpreted by a microprocessor must be converted to numerical codes before processing can take place. These codes are related to the states of transistors. The *off* position, for example, can be numerically represented by a zero, and the *on* position, by a one.

The state of a transistor is called a *bit*, which is an acronym for the term binary digit. *Binary* refers to a mathematical system that only uses two numerals (zero and one) to express numerical values.

A code created by the states of a group of 8 bits is called a *byte*.

Codes can represent letters, numbers, or symbols as defined by the *American Standard Code for Information Interchange*, known as ASCII, in addition to other information.

Software hides the complexity of codes from most users. When we type the capital letter A, for example, the microprocessor interprets our typing as the 8-digit number 0100001.

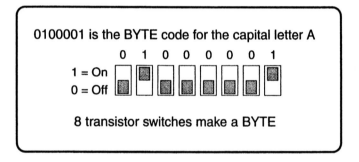

The number of bytes a microprocessor can send or receive is referred to as I/O (In/Out). While a microprocessor can simultaneously process billions of bytes, *its I/O manages only a few bytes at a time.*

You notice I/O limitations when you start up your software.

The waiting you experience at start-up is caused by the microprocessor laboriously pulling in new information. When enough information is collected inside the microprocessor, your software begins running. When the microprocessor requests new information, or it writes processed information to your disk, you experience these pauses again.

At this writing, personal computers use 32-bit chips. This means that they have the I/O capacity of 4-bytes at a time. Slower chip-types, like those once used in older computers, are now used in small, independent EAC devices with limited processing needs. These include 16-bit, 8-bit and even 1-bit chips.

In the early 1980s, the largest chunk of information an 8-bit microprocessor could process was 65,536 bytes. When data required more than 65,536 bytes, it was broken up into *subsets* (smaller chunks) that would fit into the processor's memory. This increased the number of steps required for processing.

A 32-bit microprocessor can process 4,294,967,296 bytes (4+ billion!) at one time. Consequently, software designed for 32-bit microprocessors will not work on old 16-bit computers because it does not break information down into small enough subsets.

The microprocessors found in workstations and mainframes have much more I/O and can process a huge number of bytes without breaking this data into smaller subsets. These microprocessors are difficult to manufacture and are extremely expensive compared to those found in personal computers.

Personal computers, however, are becoming more powerful. The 64-bit microprocessor, for example, was at one time only used in expensive engineering workstations. It is predicted that within a few years, this fast chip will become the norm, completely replacing 32-bit technology.

What does 64-bit technology mean to you?

With the exception of communications, which can significantly narrow the number of BYTES that flow through cables, you will rarely experience system slowdowns during normal processing. Information will flow so quickly, it will almost seem alive.

Parallel processing was developed to increase processing capacity, regardless of microprocessor type, without significantly increasing costs. It uses two or more standard microprocessor chips on the same bus (circuit board), thereby increasing I/O and processing capacity without the cost of using expensive chips, such as the 64-bit chip.

The *massively parallel computer* is a mainframe computer that uses numerous, somewhat common microprocessors. This technology has significantly reduced the cost computers used for scientific and financial applications and has almost stopped the development of extremely expensive supercomputer chips.

No matter how many bits a microprocessor may contain, I/O is dependent on the exchange of information through circuit boards, as discussed in the next section.

Circuits and Circuit Boards

Electrical circuits route electricity to and from a variety of sources. A circuit board increases the number of routes available within a small area by providing a stiff backing upon which ultra-fine wiring and multiple connection points rest.

Microprocessors and other electronic devices are plugged into specially designed circuit boards. At the edges of these boards are *terminals* (also known as connection points or plug sockets) which provide the means to attach wires from other devices. These terminals are usually exposed behind your computer's case.

Computer circuit boards also contain sockets used to connect additional circuit boards. These boards are commonly called *"add-on boards"* or *"cards"* and sit perpendicular to the main board. Installing a new card is sometimes referred to as *"taking up a slot."*

The term *bus* refers to a *group of connections* on the main circuit card that move data between the integrated circuit chips. The type of bus used defines the types of cards that can be plugged in and **directs the flow of information through wires** in much the same way as a freeway system directs drivers to use traffic lanes.

A 32-bit microprocessor requires a group of 32 wires, called a 32-bit path, on the bus. Other devices use fewer wires, which can cause slowdowns. In a freeway system, switching from many lanes to just a few causes backups when traffic is heavy. The same is true with the flow of information. Crossover points between different wiring requirements on the bus slow the flow of data, even though a fast computer chip is being used.

The following list provides a few examples of common bus names and their wiring capabilities:

8-bit bus:	Sends data along 8 wires.
16-bit or ISA bus:	Sends data over 8 or 16 wires.
EISA bus:	Sends data over 8, 16, or 32 wires.
MCA bus:	Limited to sending data over all 32 wires.
Local or PCI bus:	A special 32 wire bus that can transfer data at 32 bits to all high speed connections, such as used for the display or disk controllers and fewer bits to other devices, such as the keyboard and mouse, which do not require data transmission speed.

Port Types:

In the process of setting up your computer, you will come upon references to various connection types, which sometimes can be confusing. Here are some examples:

A *port* is another name for a connection.

The names *serial port* and *com port* are used interchangeably. The abbreviation "com" stands for "communications." They provide the connection that allows information to be sent or received one-bit-at-a-time to another device or outside source. These ports are commonly used for modems and mice, both of which communicate one-bit-at-a-time through a 9-pin connector, although older computers and a few off-brands use 25-pin connectors.

The names *parallel port* and *LPT port* are also interchangeable. LPT used to stand for *line printer terminal* because at one time, only printers were connected to computers in a parallel fashion. This type of connector has 25 pins, which are capable of simultaneously sending or receiving information over 16 wires, 8 bits at a time.

Memory

It is extremely important that you understand what memory means, even if you don't understand all the acronyms associated with the subject.

> *Memory* refers to any type of *media* (electronic or material) that stores information for eventual use by a microprocessor.

The media used for magnetic stripe cards in an EAC system is made of the same material as a computer's floppy disk. It requires special readers to move information to and from the computer, thus slowing the rate of information transfer. Transistor technology, however, is used to build microprocessors and memory chips. It transfers information between chips quickly because it does not require a "middle-man" reader or a mechanical process.

Memory has three important characteristics:

1. **The length of time it can securely retain information.** The longer it can retain information under the greatest range of circumstances, the more secure it is.

2. **The speed at which it can release or acquire information.** The faster the speed, the more efficient it is.

3. **The compatibility of the media with the current system.** When a compatible system is available, information can be retrieved. Note that when a system is changed, it is often necessary to convert stored information to a new format. If this isn't done, information becomes permanently unavailable even though it is still on the media.

Electronic memory releases and acquires information quickly, but does not retain information when the power is off. The exception to this is a product called *flash-memory,* which is simply RAM backed up by a battery.

Other types of memory, such as found in hard and floppy disks, hold information for a long time, but because they are controlled mechanically, they release and acquire information slowly compared to electronic processes.

The microprocessor works fastest when it gets information from RAM. In this case, all the chips involved transfer information electronically through transistor technology.

Although powerful, the microprocessor does not work when it has no information. Consequently, it pauses when acquiring new information until enough is present to begin processing. This pausing is noticeable when the fast microprocessor acquires information from or stores information to a slow mechanical hard drive.

The Way RAM Works

Due to the speed of information exchange, the more RAM you have, the faster your computer works because the computer can get the needed information from RAM faster than from any other source.

RAM holds a temporary *copy* of information that sits on your hard drive or floppy disk. This means that during processing, your information is often in at least two places: on the hard drive and in RAM.

Electronic Memory Chips

Three very important types of transistor-based chips used in computers and intelligent devices are:

BIOS chip

BIOS stands for _Basic_ _In_ / _Out_ _System_. It controls information going *in to* the microprocessor, as well as that going *out* to other devices.

BIOS is the first electronic memory chip activated when a system is turned on. It contains encoded instructions, which it sends to the microprocessor. These instructions include information about the computer's components and their configurations.

BIOS chips require very little energy and are backed up by small batteries so that they do not lose their information when power is off.

Microprocessor chip

This chip receives, holds or sends information as well as performs calculations on the information it contains.

DRAM chips

These chips, which are commonly called "RAM chips," provide a *volatile,* ever-changing memory. They acquire information by *copying existing data* from magnetic or optical hard drives, CD-ROMs, etc., or by receiving new information from the microprocessor.

DRAM chips exchange information at speed of millions of bytes per second (megabytes), making them highly efficient. The more information in DRAM can hold, the faster the microprocessor appears to work because it doesn't have to wait long for information transfer.

Acronyms for Memory

Acronyms for computer technology are being made up daily. Many times manufacturers invent them to help create differences between products, thereby making their products sound special.

Specification sheets on many security-related devices used in EAC contain acronyms, consequently, it is wise to stay up-to-date on usage.

The following list contains six common acronyms that refer to memory:

RAM or DRAM   *Dynamic Random Access Memory* - this memory can be accessed "at random" and is not controlled linearly, such as the information found on a straight magnetic tape.

ROM   *Read Only Memory* - this type of memory cannot be changed once it is recorded. A CD-ROM, then, refers to a compact disk (CD) that contains prerecorded information that cannot be changed.

PROM   *Programmable Read Only Memory* - this type of electronic memory behaves as a ROM once it is encoded. It is a separate chip that generally contains the instructions used by the microprocessor.

EPROM   *Erasable Programmable Read Only Memory* - this type of memory is meant to function as a PROM. Under specific circumstances, it can be erased and reprogrammed. It is commonly used during the development of a new microprocessor design.

Chapter 6

Variations in Acronyms

The variations of acronyms using "RAM" makes a very long list. Consequently, it is important for you to know how to interpret the "root words" of acronyms (which can change on the whim of a manufacturer) in a way that makes sense to you.

The following list will give you an idea on how RAM-related acronyms are used:

CDRAM	cached DRAM (cached means "another layer of memory")
CVRAM	cached VRAM
DRAM	dynamic RAM (dynamic means that the "memory is constantly refreshed")
EDRAM	enhanced DRAM
EDORAM	extended data out RAM
EDOSRAM	extended data out SRAM
EDOVRAM	extended data out VRAM
FRAM	ferroelectric RAM
RDRAM	Rambus DRAM
SDRAM	synchronous DRAM
SRAM	static RAM (static means that the chip "holds information as long as power is present")
SVRAM	synchronous VRAM
3D RAM	Matsushita's chip for 3-D video processing
VRAM	video RAM which buffers the monitor refresh rate
WRAM	window RAM

Electronic Speed

The power of a microprocessor (chart on next page) is determined by the:

- Speed at which it can acquire information from other sources, such as electronic, magnetic and optical memory, as well as from other devices (keyboards, mice, etc.). This is referred to as I/O (in and out).

- Clock speed, in megahertz, at which the microprocessor's internal switches move during processing.

As mentioned earlier, a 32-bit microprocessor is faster than a 16-bit microprocessor because more information can flow to and from it and it has greater information holding capacity (more transistors).

The term *clock speed* is used to designate differences within similar microprocessors. It refers to the shortest time under a second in which any operation can happen. Simply put, the faster the microprocessor, the faster its internal switches move.

Clock speed is measured in *megahertz* because switching currently can be done millions (mega) of times per second (Hertz). This speed is sure to increase in the future as new switching technology becomes available.

Despite fast clock speeds, not all software performs quickly, even on the most up-to-date microprocessor. This is because software controls many devices, such as the disk drive and printer, that do not have fast I/Os. The more demand made on I/O, the slower the performance, consequently, up-to-date busses must complement microprocessor technology, as discussed earlier in this chapter.

High clock speeds switch transistors quickly. Unfortunately, physical wire becomes extremely hot when it handles fast electrical vibrations caused by processing speeds. This is a serious problem for the microscopically thin wire embedded on a microprocessor chip. Fast microprocessors can become so hot that they cause damage to the computer and on occasion, even start a fire. Fast laptop computers can actually get so hot that they become uncomfortable to touch.

Warning: Excessive heat can cause a computer to act strangely, resulting in software that periodically slows down or actually locks up the system.

To control heat, tiny fans, often smaller than 1 inch, sit on top of microprocessors to blow a continual stream of cooling air during processing. Additional fans are placed throughout computer component box, as needed. Heat

Chapter 6

generation is such a serious problem that blocking vents anywhere on a system could cause electronic problems and could possibly lead to a fire.

Preventive Maintenance Tip: When you do preventive maintenance checks, open the computer's case and check whether all the system fans are working properly.

Microprocessor Speed Improvements: The following chart, based on information supplied by the Intel Corporation, shows microprocessor speed improvements since 1978.

The speed benchmark in MIPS (millions of instructions per second), as seen on the chart, was dropped with the introduction of the Pentium 166 MHz and a new benchmark has been substituted. Consequently, we cannot show the latest models on the chart below because there are no comparisons. It should, however, give you an idea of improvements, especially after 1993.

Microprocessor Speed Improvements

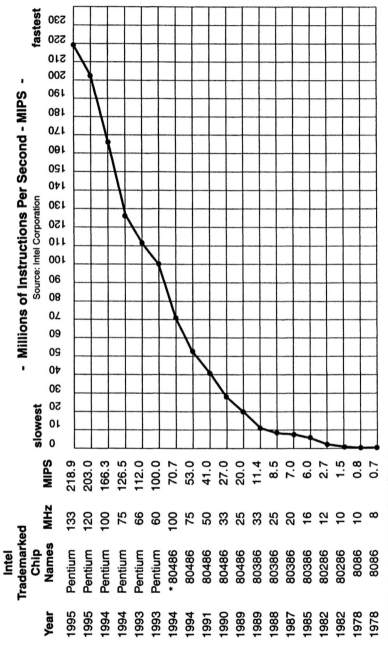

Year	Intel Trademarked Chip Names	MHz	MIPS
1995	Pentium	133	218.9
1995	Pentium	120	203.0
1994	Pentium	100	166.3
1994	Pentium	75	126.5
1993	Pentium	66	112.0
1993	Pentium	60	100.0
1994	* 80486	100	70.7
1994	80486	75	53.0
1991	80486	50	41.0
1990	80486	33	27.0
1989	80486	25	20.0
1989	80386	33	11.4
1988	80386	25	8.5
1987	80386	20	7.0
1985	80386	16	6.0
1982	80286	12	2.7
1982	80286	10	1.5
1978	8086	10	0.8
1978	8086	8	0.7

Source: Intel Corporation

* NOTE: Pentium is a trademarked name by the Intel Corporation.

Microprocessor Names

A microprocessor is a patented miniature circuit that manages 8, 16, 32, or more incoming or outgoing bits, with many more internal bits (transistors) that handle the actual processing.

Although microprocessors of various designs can accomplish the same tasks, the Intel Corporation patented a range of microprocessors that have sparked the growth of the computer industry, thereby creating the standards that others copy.

Until the release of the Pentium chip, Intel designated their chips by part number. Other chip manufacturers, as well as the builders of desktop systems, refer to the Intel names as a way of designating performance, even though Intel discourages this practice. These names are as follows:

Chip Name:	Chip Type in Increasing Power:
8086	8-bit chip (also called an XT by IBM)
80286	16-bit chip (also called an AT by IBM)
80386	16-bit chip (also called a 386)
80486DX	32-bit chip (also called a 486)
Pentium	32-bit chip (also called a P-chip)
Pentium Pro	32-bit chip that's even faster . . .
Pentium P6	64-bit chip (at one time used only for expensive engineering workstations but are now being introduced to the general public)

Although there are variations, computers are commonly identified through three system components, which are the:

- Name of the microprocessor,
- Clock speed of the processor in megahertz (MHz) and
- Capacity of electronic memory available, usually expressed in the millions of bytes (megabytes) of RAM.

Using the above identification system, the following is an example of a computer type:

Pentium Pro - 200 MHz system with 32 megabytes of RAM

It is important to make a distinction between the brand name of a computer — such as Gateway, Dell, IBM, Packard Bell, etc. — and the *type* of system it is. Brand names alone do not convey functional information about a computer.

Intelligent and Dumb Devices

An intelligent device is one that is controlled by its own microprocessor.

At minimum, an intelligent device requires its own operating system, BIOS and instructions. It can also have DRAM, Flash Memory and other means of storing information.

While state-of-the-art, superfast microprocessors might drive the computer, intelligent devices often use slower, less expensive microprocessors, such as those defined by 8-bit and 16-bit chips.

A dumb electronic device operates in a manner similar to an intelligent device, except that it is 100% dependent on another computer for instructions.

Dumb devices place an extra burden on a supervisory computer's capabilities and will fail to work properly when that computer goes down.

Overburdened computers (those that must control numerous dumb devices) send instructions slowly, severely limiting the effectiveness of a dumb device's ability to respond to emergencies.

The ideal devices, then, tend to be intelligent and can work independently of a computer, even though they might use a central computer for its:

- User interface

- Communications ability (information processing and routing)

- Information storage capability

Intelligent devices are more useful, under a wider variety of circumstances, than dumb devices. This is a very important consideration when selecting devices for a security system.

Systems - Networked and Distributed

An electronic access control network consists of one or more personal computers, control panels and devices, such as electronic authenticators, locks, sensors, etc.

Network control is provided by *network management software* that resides on the supervisory computer. The supervisory computer in a network is commonly referred to as a *server* because it deals out information.

The interconnection and exchange of information between devices in a computer network is referred to as *communications.*

For further information on networks, see Chapter 7, *Communications* and Chapter 8, *System Design.*

Distributed System

A distributed system is made up of intelligent control panels that communicate with the supervisory computer. This means that each control panel acts like an independent computer, even though monitors and keyboards are not present. The supervisory computer and control panels in this type of system communicate with one another as needed. Communication, however, is not always required.

> *Example:* An intelligent control panel might not report common activities, such as granting access to authorized users, but will report attempts at unauthorized use as well as all alarms.

Intelligent control panels stay up-to-date through software changes, just like personal computers do. In the case of improvements in technology, many intelligent control panels can also have their ROM or EPROM chips replaced. Chip replacement is an economical way to upgrade, especially when compared to replacing an entire system, which might involve rewiring!

Numerous intelligent control panels can be networked within a distributed system, sometimes using repeaters to boost signals between them and the server. All the panels, of course, work independently from one another and none put a strain on the main server.

In the past, dumb control panels were used to control access. This was primarily because of the expense of the microprocessors involved.

Fortunately, microprocessor and memory prices have dropped. Distributed systems and intelligent devices are becoming the standard. This enables EAC systems to stay running no matter what problems may befall any part of them—most certainly a plus for security!

For further information on distributed systems, see Chapter 8, *System Design.*

Chapter Review - Computers

The three main types of computers used in a large electronic access control (EAC) network are the:

1. Controllers (or *Control Panels*) that control devices such as authenticators, locks, gates, etc. at the site.

2. Switchers that control closed circuit TV cameras and video taping activities.

3. Supervisory computer that oversees the entire network, including the other computers.

It is important that you identify all your hookups and switches because . . .

this keeps others on your staff well informed and prevents fumbling in the case of an emergency.

Building redundant systems is important because . . .

duplicating systems provides backups in the case of emergency and increases the probability that your facility will be protected in all situations.

EAC software is often regarded as database-type software. To keep it running smoothly:

1. Plug your computer into an *uninterruptable power supply* to make sure power problems don't shut down your system.

2. Post troubleshooting procedures before a problem happens.

3. **Most important:** Make daily backups of all files.

Your computer's *bus* (circuit board) is important because . . .

it channels electrical impulses from one of many sources to the microprocessor and then from the microprocessor to one or more destinations. The bus should have as many paths as possible to keep data transfer fast and smooth.

The word *port* means . . .

you are referring to a physical connection point on a computer.

Electronic memory (RAM) and microprocessors are very closely related because . . .

> they transfer information electronically, based on transistor technology. Information stored in RAM can be fed to the microprocessor almost as fast as the microprocessor can process it.

Three important characteristics of usable memory are:

1. The length of time it can securely retain information.

2. The speed at which it can release or acquire information.

3. The compatibility of the memory media with the current system.

Acronyms use the initials of words to create new words. How should you approach unfamiliar acronyms?

> When you see an acronym, write it and its meaning down. Keep adding to your list so that you can stay up-to-date. Once you understand how "root words" are used in acronyms, you will be better able to grasp new technical terms.

The differences between RAM, ROM and PROM are:

> RAM refers to an electronic memory chip based on transistor technology. The information it holds can be changed instantly; however, when power is disconnected, it loses that information.

> ROM refers to a memory vehicle (which could be a chip) that contains information that cannot be changed. Further, this information is unaffected by the loss of power.

> PROM refers to a memory chip based on transistor technology that can be changed by using special technology. Once it is changed, it behaves like a ROM. PROMs are commonly used in control panels and other computerized devices. They can be upgraded to significantly improve performance without requiring the purchase of new hardware.

Electronic speed affects computer power by. . .

> increasing the speed at which information is transferred between devices and by increasing the heat generated by the computer components.

Good preventive maintenance includes . . .

> checking to see that all the fans are working inside the computer component box and making sure that vents are clear in order to keep your system cool.

Computer types are usually described by the . . .

- Generic name of the microprocessor.

- Clock speed of the processor in megahertz (MHz).

- Capacity of electronic memory (RAM) in megabytes.

An intelligent system is one that . . .

> contains computerized devices that function without the controlling input from a supervisory computer. If an intelligent component of the system fails, it does not affect the performance of other devices on that system. If the supervisory computer fails, the remaining devices continue to work.

A distributed system is one that is made up of . . .

> intelligent devices.

QUESTIONS

1. Why is it important to identify the components in your computer?

2. In an EAC network, what three types of computers are most commonly used?

3. Why is it important to use intelligent devices?

4. The computer bus (or main circuit board) is extremely important. What bus characteristics must you look for when purchasing a new computer?

5. Why is it that all memory, whether it is RAM, a hard drive or an EAC credential of some type (such as a magnetic stripe card), performs basically the same function? What is that function?

Chapter 7

Communications
Wired and Wireless

Connections

Security professionals are spending more and more time peering into monitors and less time "on the beat." This is because an electronic access control (EAC) system is a large computer network. Consequently, understanding hardware, especially the connections that make computers work, is becoming increasingly important.

The maze of connections associated with computer networks is confusing. Without planning and organization, the simple prayer "we hope nothing goes wrong" takes on sinister meanings. If the security force can't quickly correct a connection problem, the premises are left unprotected.

Full-time technicians are familiar with connection principles. They earned this knowledge through special courses and on-the-job experience. Most other people are not.

Fortunately for nontechnical people is the fact that if you organize your information about connections, a little knowledge goes a long way.

Quick Technical Education

The best way to acquire a quick technical education about connectivity is to buy books and magazines that provide numerous, colorful, clear illustrations. You need to see how a connection is made in order for you to properly understand it.

In the security industry, the very best publication of this type is the quarterly *SDM Field Guide*. Consider purchasing back issues to build an excellent reference set.

How Networks Work, published by Ziff-Davis Press, illustrates how computer networks work. It includes information about telephone systems, modems, cabling and networking types.

The National Burglar & Fire Alarm Association (NBFAA) sponsors the National Training School (NTS), offering comprehensive training for alarm company owners, technicians, central station operators, salespeople and administrative staff. Membership in this organization requires a financial commitment, which pays off by focusing your attention on exactly what you need to know when you must know it. (See *Appendix* for address.)

Fast Study: Last, if you are short of study time, but are required to quickly learn a lot of new technical information that's written in articles or documentation, do the following:

1. Using a highlighting marker, mark key ideas throughout the material.

2. Rapidly read this information out loud into a tape recorder. You do not need to sound like an announcer and don't mind if you fumble.

3. Listen to this tape several times when you drive or when you might otherwise be listening to the radio.

4. If there is a part you must memorize, talk along with the tape recording, duplicating what you hear.

Connection Inventory

Treat connection points with the same professional organization used to maintain ID information.

A well-organized connection inventory reduces maintenance costs by providing the means to keep technicians well-informed. Most important, it provides insurance against losing valuable knowledge when "the last senior person" quits or retires.

Organize your information. Know the exact location of each device, its connection points and technical details.

The following list pinpoints the type of information that should be kept on individual devices:

> **Technical Inventory of Equipment:**
> Type of device
> Manufacturer
> Installer, if any
> Physical requirements, including plug types, wires and/or cables
> Voltage requirements
> Locations
> Wiring paths and/or junctions (connection points)
> Software name, if any, and most recent version installed
> Cautions and warnings

Every change to your system requires documentation. This can be an easy task that does not require reams of paper and hours of typing when you know how to use technology.

A **database** kept on a computer can provide an efficient way to manage this type of information. It can, however, be maintained on handwritten **index cards**.

Take **photos** of the installation sites and wiring connections. Enlarge them to 5" by 7" so that their details are clear.

File these photos, along with product literature and instruction sheets. If you can use a graphical database on your computer, scan this information so that the pictures and illustrations are present whenever information is seen on your monitor.

Video recorders with sound are inexpensive (as low as $400) and produce tapes that are good enough for inventory purposes. If lighting is dim, use a high-powered flashlight during filming. Make tapes of your devices and systems, describing them as you go.

Use *Snappy*, a $200 video frame grabber that attaches to your computer's printer port, to capture pictures from your video recorder. Once "snapped," you can store these images in computer files and/or print them on paper to be stored with your standard files.

Digital cameras are also inexpensive, however, you'll find that a **video recorder does a better job** of capturing your system and maintenance procedures. Video cameras "see" more information than single shots and often include detail that you might have forgotten about.

Another plus to using video recorders is being able to **verbally describe situations** while you record them. This method often provokes deeper reflection on the subject than you might have when writing a report, especially if you are not used to writing and time is tight!

Facility Maps

The placement of security devices and connection points throughout a facility can be easily seen on facility maps. These maps show the placement of devices and wiring paths through the use of symbols and can be as detailed or simple as you wish.

Excellent examples of color-coded facility maps can be seen in the *SDM Field Guide*. The symbols they use are available on a poster that costs $5, including shipping.

> *Contact:* "Symbols Poster," Security Distributing and Marketing, PO Box 5080, Des Plaines, IL 60017-5080.

Information Updates and Training Tapes

Update the information about your site on a regular basis.

If you use a video recorder to do this, you'll find yourself making training tapes for new staff at the same time — extremely cost efficient!

Video tapes make a wonderful way for senior staff (possibly people who are about to retire) to pass their knowledge to the rest of the security team.

Keep All Your Information Secure

Obviously, good record keeping procedures simplify the maintenance of an EAC system. All records must be kept secure, however, so that they do not fall into the wrong hands.

Establish good procedures so that you can always identify who is using your materials. Regularly check your files to make sure that everything is in place.

Wiring Audits

Security managers are usually responsible for all connections related to their area of command. Regularly scheduled wiring audits let them check whether the:

- wiring is healthy (not corroded),

- connections are solid,

- connections meet code and

- system has been exposed to tampering.

Normally, security personnel do not have the expertise to install systems. This fact should not, however, make them overly dependent on outside installers and consultants, or limit the amount that they can learn about their own installation.

To make sure that connections are made to specification, check the credentials of every member of the installation team. Next, audit the installers' work, both in progress and when complete. Take nothing for granted. Hire a third-party expert and accompany this person during the audit process. Learn as you go.

Take photos or video tapes during the audit. Use these to train your staff as well as to serve as a reference as to how connections should look, so you can tell when they are broken.

> **Tip!** Taking a video tape during an audit allows you to examine what was said later in a more relaxed atmosphere. Use the video to train yourself in technical matters, then pass it to future managers so that they know what to look for.

Cable Jackets

Everyone knows what a very simple electrical extension cord looks like. Not everyone knows, however, that the two prongs on the plug-end indicate that there are two wires inside the cable, each serving a different function.

In general, every prong, pin or pinhole on a plug is attached to a wire housed inside a cable. Even tiny connectors, such as telephone jacks, are attached to several wires. These unseen wires hide the means of connectivity from us, making the connection process seem mysterious.

If you split a cable open, you'll often notice that individual strands of wire are wrapped in colorful coatings (called *"jackets"*). Manufacturers use colors to code the wire for easy reference.

Wiring Practices
National Fire Protection Association
NFPA 79 - Electrical Standard for Industrial Machinery

Wire Color	Description
Black	Line, load and control circuits at line voltage.
Red	AC control circuits, at less than line voltage.
Blue	DC control circuits.
Yellow	Interlock control circuits supplied from an external power source. (International standards=orange)
Green	Equipment grounding conductor where insulated or covered. May have yellow stripes. (International standards=green and yellow)
White	Grounded circuit conductor. May also be natural gray. (International standards=light blue)

Both cable and individual wires are wrapped with a variety of materials, including those final jackets. These wrappings include insulation and shielding materials, as needed, and sometimes strengthening materials that protect the wire assembly against stretching during installation. The materials used are selected for specific applications, such as for exterior and interior installations, protecting against dampness, cold and heat and shielding from electromagnetic interference.

Cable assemblies must be flexible enough to let all included materials expand or contract at different rates. Consequently, the thickness of a cable assembly is determined by the material needs of all its parts.

Cable Types

The term "conductor" refers to metallic material that carries electrical current. Nonconductive material, such as the glass used in fiber optics, does not carry current.

The illustration below shows a sample of conductor cables.

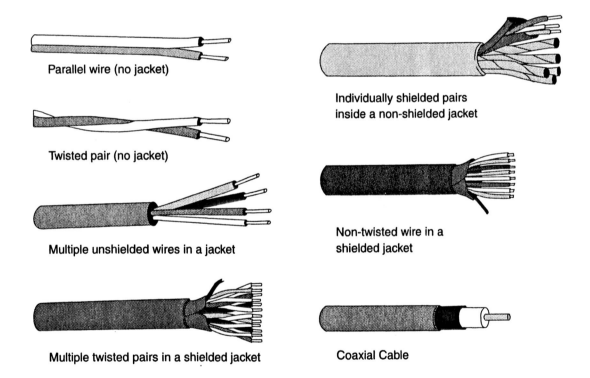

Parallel wire (no jacket)

Twisted pair (no jacket)

Multiple unshielded wires in a jacket

Multiple twisted pairs in a shielded jacket

Individually shielded pairs inside a non-shielded jacket

Non-twisted wire in a shielded jacket

Coaxial Cable

Conductor cables contain copper wires and require a ground wire or grounding material within the jacket, in addition to the wires that carry current. Nonconductive material (fiber optics) does not require a ground.

The three types of conductor cables commonly used in communications are:

Unshielded Twisted Pair (UTP):

A circuit requires two wires; one for sending and the other for receiving. Each wire is covered by a jacket. Typically, the cable used in communications holds four pairs of twisted wire. Each pair is twisted differently, with the number of *twists per inch* being the defining factor.

Twisting each pair of wires helps cancel *noise* (electrical signal interference) from the adjacent wires within the cable as well as from other devices in the building, such as motors, relays and transformers.

Shielded Twisted Pair (STP):

This cable holds two sets of twisted wires. Each set is wrapped in a foil jacket. Both foil jackets are wrapped together inside a braided copper mesh and then the whole assembly is wrapped by an outer jacket. This cable is used for some computer networks, such as a *Token-Ring LAN*.

Coaxial Cable:

This cable is commonly used for video hookups as well as various types of computer networks. It has a single copper core, with the second conductor being the shield that surrounds the core. The core is packed in plastic insulation, which is then wrapped in the shield (braided copper mesh) and is finally covered by an outer jacket. The connector is tubelike and has only one pin.

Fiber optics provide nonconductive cabling for communications:

Fiber Optics:

Fiber optic cable is made up of glass or plastic *filaments* (slim threads) that allow the transmission of light. Light transmission over each filament creates a stream of bright and dark spots, which, when measured over time, result in an ON/OFF code that can be interpreted.

There are three primary advantages for using fiber optic cable:

First is that fiber optics are totally free from electrical interference.

Second is that they can carry data further without signal degradation. Fiber optics covers more than 11 times the maximum distance for coaxial and 15 times the distance for twisted pair cable before signal boosters are required.

Third is that light transfer can be precisely controlled, almost eliminating the possibility of tapping. Conductive cabling can be tapped, allowing the eavesdropper to read all passing data, including unencrypted passwords. In a precisely adjusted fiber optic situation, however, tapping alters light transmission patterns, causing the entire system to fail before information is illegally captured. This makes fiber optics highly tamper resistant.

Fiber Optic Unit

The cladding and core are fused together.

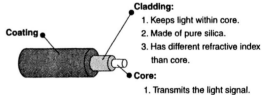

Cladding:
1. Keeps light within core.
2. Made of pure silica.
3. Has different refractive index than core.

Core:
1. Transmits the light signal.
2. Made of Germania Doped Silica.

Coating

Inside View
Fiber Optic Breakout Cable

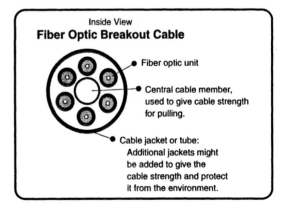

Fiber optic unit

Central cable member, used to give cable strength for pulling.

Cable jacket or tube: Additional jackets might be added to give the cable strength and protect it from the environment.

Conduit Piping

As everyone knows from looking at the back of a PC, the profusion of individual cables can be messy. To avoid tangles, cables are threaded inside a *conduit pipe*, which is a hollow metal or plastic tube that is run between walls.

> *Security Recommendation:* To prevent tampering, all security wiring should run in unmarked conduit and not in loose raceways (i.e., general areas where various cables can run). This includes all wiring from the devices to EAC control panels and from the control panels to the main computer.

Conduit categories are based on various smoke and flame characteristics. The best grade, which is also the most expensive, is General Purpose. It can be used in any of the four conduit categories.

Conduit Categories:

1. Residential (CM-X) — lowest grade
2. General Purpose (CM or CM-G) — commercial grade
3. Riser-rated (CM-R) — runs up or down walls between floor levels
4. Plenum (CM-P) — runs in ducts used for environmental air as well as air spaces in the ceiling

Older facilities may not have enough spaces available to hide conduit. In this case, wires can be run though a *square channel raceway* attached to the exposed walls. This is not, however, a recommended method as it greatly heightens exposure to tampering.

Cable is pulled through conduit, which puts a tremendous strain on the cable assembly. Consequently, the sturdiness of cable is rated by its maximum pulling tension and care must be taken that the cable used will stand up to installation, in addition to providing the right connections.

Cable Pulling Factors That Determine Tension Rating:
Conduit fill ratio (number of cables to conduit)
Friction
Number of bends (corners) required
Conduit material
Cable jacket material
Maximum cable tension before snapping
Amount and type of lubricants used to ease friction

Some cable jackets, like those used in fiber optics, contain materials that are specifically designed to take the entire force of the pull without stretching or cracking the contents. Special tools are required to make use of this jacket-type.

Care must be taken to not yank a cable through a tight conduit packed with other cables, or around a sharp corner, as this can break wires. Once the cable is installed, breaks are difficult to pinpoint and are expensive to fix.

Needless to say, it is important that cable be repeatedly tested for wholeness throughout the installation process, or nasty surprises can result at the end.

Cable Specifications and Splicing

Most installation guides specify cable and connector types. They also tell you the purposes for which individual wires within the cable are used and how those wires attach to specific prongs, pins, or pinholes on connectors. If all this information is not packaged with the device, you can request it from the manufacturer.

The wires bundled within one cable can be transferred to several different cables through the process of *splitting*. To split a cable, you must first remove that cable's jacket far enough to expose the enclosed wires, then *fan* (separate) the wires for easy handling.

Fanned wires are joined to different wires through *splicing*. This requires that the jackets be removed from the wires so that the physical wiring material is exposed. This material is connected to the same material in a second wire through a number of methods, the most common of which employs a connector cap that clamps the two together.

Connections might be made device-to-device or made in special junction boxes (which are also known as *"breakout boxes"*). These boxes protect spliced connections and usually serve as rerouting centers for joined cable.

Conduit is rated as to whether it can hold:

- Conductive wire (contains metallic properties)
- Nonconductive wire (contains no metallic properties)
- Composite wire (contains metallic and nonmetallic properties)

They are also rated for maximum voltage circuits. Grounding is required for all conductive and composite wires in conduit and in junction boxes.

The flow of information deteriorates over long wiring paths. To boost the information signals, repeaters are installed at intervals. These devices rebroadcast information, giving the flow a fresh start from that point.

Tips for Troubleshooting Cable Installations

- Heavy objects dropped on a cable can cause wire crimping and possible breakage.

- The bend rules for cable depend on the number of conductors and the thickness and type of conductor. Consult the manufacturer for recommendations.

  Sharp bending can cause crimping and breakage. In some cases, right angles (90-degrees) actually change the characteristics of the cable, resulting in reduced data transmission rates. Bends with a 3-inch radius are acceptable for most coaxial cable.

- When fiber optic cable, which consists of glass or plastic filaments, is bent or yanked through tight areas, filaments randomly snap apart. If fiber optic cable is bent at a right angle, it can't transmit light, even if the cable itself is perfect.

- When the joint between the wires and the connector (plug or prong housing), is bent sharply, it can cause random breakage.

  Remember: Cables carry many wires, each used for different purposes. A single broken wire inside a cable might let a device work, but will cause intermittent problems.

- The function of one type of cable can interfere with another.

  This problem comes under the heading of *EMI* (Electromagnetic Interference), where strong electromagnetic signals interfere with digital signals.

  An installer can render the best EAC system useless if EMI is not considered.

  To avoid this, a good practice is to isolate communications wiring from power wiring, putting each in its own conduit.

  The use of high shielding conduit also solves many problems, especially if the communications line passes through areas filled with electromagnetic devices, such as electric motors and even ballasts for fluorescent lights.

  Fiber optic cabling, of course, is not affected by EMI at all; however, it might be impossible to use if the installation path requires many sharp bends, a tight lay in conduit, or bends of 90-degrees or more. These conditions can cause breaks in glass filaments or block signal transfer, rendering the cable useless and troubleshooting difficult.

- Problems occur when data cables cross power cables. In this case, make sure that the cables cross at right angles, to minimize exposure.

- The function of other devices, the physical environment and/or weather interferes with the installation.

  Check your building and surrounding buildings for machinery and electromagnetic device usage that may cause EMI problems or power drains. Check local storm patterns and geographical particulars (such as mountains and iron ore mines) that are known to cause problems. Also talk with neighbors to determine if they had any problems due to location, power utility service, etc.

- There is not enough power to send the data signal over a long length of wire.

  This situation requires the addition of *repeaters*. These rebroadcast information from points just prior to where the loss of signal strength becomes serious.

- The power that is provided is not clean.

  Before installation, always perform a power check on the circuit for spikes, anomalies and ground. The results of this check will determine what you need to do to protect your system from power problems that could render it unstable and unreliable.

  Invest in *UPSs* (uninterrupted power sources) that clean the flow of power and make sure everything is properly grounded. Common *surge protectors* may not provide enough protection for a security system.

- Terminal ends (where the wire meets a connection point) are tightened with too much pressure, causing weakness and potential breaks.

- Terminal ends are not tightened enough, causing poor connections and potential separation.

- The diameter of the conduit is not large enough to hold the required cables to allow for jacket swelling during climate changes or manipulation of existing cables when repairs are required.

- Everything seems to be OK, but the equipment doesn't work.

This may seem obvious, but it is often overlooked. *Use the manufacturer's cabling specifications.* Substitutions can cause terrible problems. In addition, make sure that the contractor does not use bits of non-specified cable to make up for shortfalls.

● **Related Troubleshooting and/or Planning Tips:**

See "Transmission Considerations" under the *Radio Frequencies* section of this chapter.

Also see the *Transmission Considerations* section at the end of this chapter, including "Problems with Signal Carriers and Receivers" and "Transmission Consideration Summary" subsections.

Wireless Connections

As of the mid-1990s, the development of wireless applications has been fueling a technological revolution. Here's why:

1. **Portability:** Wireless devices can be set up in temporary sites. Visitor traffic, for example, might create the need to increase security in a given area for only a short period of time. Wireless access control, motion sensors, CCTV and other devices can be installed in such a case and then dismantled without extensive carpentry or damage to the environment.

2. **Accessibility:** Wireless devices can be put where wired devices can't easily go. EAC in elevators, for example, requires long stretches of cable. Wireless EAC, however, can be installed with a minimum of fuss.

3. **Creativity:** New wireless transmission techniques are being perfected that provide greater ranges, more jam-proof signals and no need for licensing. Among these is the increased use of frequency hopping technology, better known as *spread spectrum*, that was developed by the military with high security in mind.

In 1993, revenues for in-building, wireless hardware exceeded $100-million. By the year 2000, that figure could easily reach $1.9 billion worldwide.

In 1994, six million portable notebook computers were sold, most of which required a link to desktop computers and over 50 million cordless phones were in use throughout the world.

Experts believe that the above figures mark only the beginning of a movement that may surpass the computer revolution. One reason for this is that as the technology emerges, new applications will be developed that cannot be predicted today.

Portability of communications is important, especially to users of wireless phones and cellular services. The security industry, however, has more specialized demands. It must move data (such as held by electronic access control) quickly and with few physical constraints.

Cabled data-moving systems require massive investments in cabling and switching equipment. Wireless systems, however, can bypass these hardware investments even though throughput speed might sometimes be sacrificed.

The Personal Computer Memory Card International Association (PCMCIA) developed a small circuit the size of a credit card that plugs into the back of a computer. One PCMCIA device lets anyone create his or her own wireless computer network, or else instantly link with existing networks. Best, it uses highly secure spread spectrum technology, mentioned above.

In the right environments, radio waves can penetrate objects and walls without any signal disruption. Infrared technology, which depends on line-of-sight transmission, is struggling to compete. Tried and true infrared components include LEDs, laser diodes and photodiodes.

New infrared *transceivers* (transmitters/receivers) are being developed that can bounce and reflect signals off walls and ceilings. Given the height of signal travel, these are not prone to random interruption such as may be caused by pedestrian traffic. The result is an infrared wireless network that can transmit at 100 Mbps throughout a 25 by 25 foot area. A new line of signal repeaters (power boosters) are increasing the distances of infrared transmission. Best, light technology is not subject to regulations, it is free from restrictions worldwide and it ensures data access across international boundaries.

Spread spectrum radio frequency technology, however, has captured everyone's imagination. Its secure signals, low power and general freedom from licensing makes it perfect for small and large computer networks, alike. It can be used for building-to-building transmissions and even as the backbone of new public communication services.

Currently, the technology behind wireless voice service is shifting away from analog and is moving to digital signals. Unlike voice, however, wireless data transmissions are extremely sensitive to errors caused by signal fading. Fortunately, emerging technology is boosting data signals, making it feasible to transmit data through cellular services. It is believed that the next generation of mobile radio communications terminals will be used more for data transmission than they will for voice.

Tied to these changes is the need for power conservation. This is forcing battery technology to provide longer life. It is also affecting transceiver design.

To lighten power requirements, components of transmitters are being eliminated. According to a representative at Hitachi, for example, their MOSFET modules do not need a negative supply voltage, nor do they require a drive circuit or power-supply switch and have zero current drain when the phone is not transmitting.

Europeans are ahead of the United States in terms of digital wireless transmission. As of 1994, digital transceiver trials were being conducted in Germany, England, France, and the Benelux and Nordic countries. The objective, backed with PCMCIA cards, is to provide wireless connectivity to any system anywhere around the world.

Radio Frequencies

Most, but not all, wireless transmissions are based on radio frequencies (RF).

RF signals can be plotted on an *analog chart,* which measures amplitude and wave cycles over time. Understanding the chart, however, requires knowledge about amplitude, sinusoidal waveforms, harmonic content, carrier waves and phasing, to name just a few things.

The general public is most familiar with RF signals as described in *Hertz (Hz).* This indicates the number of wave cycles (vibrations) per second that are present in the signal. These can be expressed as follows:

Numbers 1 to 999:	50 Hz, 250 Hz, 875 Hz, etc.
Thousands:	1000 Hz or 1 KHz
Millions:	1,000,000 Hz or 1 MHz
Billions:	1,000,000,000 Hz or 1 GHz

Signals are not limited to RF, however. In nature as well as man-made environments, many frequencies from differing origins compete with one another for airspace. Those that win disrupt or jam the losers.

The RF signals that control a car burglar alarm, for example, are often within the range of signals generated by an electrical storm. This results in false alarms.

Unwanted signals are a serious problem in security as well as in broadcasting. Consequently, governments have tried to control the problem by restricting users to specific frequencies within the *frequency spectrum* (range of all frequencies possible). This control is in the form of broadcast licensing and regulatory restrictions.

The range of frequencies in which a signal operates are called *frequency bandwidths* or just plain *bandwidths.* The following chart indicates the regulated bandwidth allocations in the communications industry.

Communications
Bandwidth Assignments

824 - 849 MHz 869 - 894 MHz	Public Cellular
896 - 901 MHz 930 - 931 MHz	Private Land and Mobile
901 - 902 MHz 930 - 931 MHz	Narrowband PCS "Personal Communications Services" - designed for broadcast within a complex, such as a campus. Carries voice, fax, and data.
902 - 928 MHz	Industrial Unlicensed commercial use such as cordless phones and computer networks (LANs)
931 - 932 MHz	Common Carrier Paging
932 - 935 MHz 941 - 944 MHz	Point-to-Multipoint and Point-to-Point
1.85 - 1.970 GHz 2.13 - 2.15 GHz 2.18 - 2.2 GHz	Computers and Computer Components
2.4 - 2.483 GHz	Industrial Unlicensed commercial use such as for computer networks (LANs)

All RF transmissions require antennas. At the simplest level, an RF transmission requires a transmitter, a transmitting antenna, a receiving antenna and a receiver. When a transmitter is combined with a receiver, such as in a cordless phone, it is often called a *transceiver*.

The transmitter contains the power to broadcast the signal. If the transmitter is underpowered, the signal will not reach its destination.

Transmitter ranges are enhanced by devices called *repeaters*. These receive the broadcast signals before they become weak and retransmit them at full strength. A cellular phone system, for example, is made up of a series of repeaters placed throughout a community. The messages they control hop from repeater to repeater until they get to their final destinations.

Users of radio frequencies and other electromagnetic signals (such as the flow of information between computers) need to understand what it takes to make a clear, uninterrupted signal and, conversely, how that signal can be jammed. The following list describes some considerations:

Wireless Transmission Considerations:

- *Government Regulation and Licensing:* All RF and/or high voltage devices must meet government regulations, whether or not they require licenses. If they do not meet regulations, their potential for being jammed or jamming other devices is high. Do not rely on product labeling to establish regulation conformity. When buying security equipment, double check manufacturer specifications directly with the FCC.

- *Surrounding Building Materials:* Metal substructures to buildings can block and/or distort signals; some metals (like brass) more than others. If the RF device is portable, check for metals in all of the areas in which it will be used.

- *Surrounding Natural Environment:* metallic particles found in surrounding stone and/or earth can block and/or distort signals. RF transmission of any kind is difficult, for example, in communities that are rich in iron ore.

- *Weather Changes:* Check the weather change patterns in your community. Periods of heightened electromagnetic activity found in lightning storms can block and/or distort signals, whether transmitted through RF or cable.

- *Surrounding Equipment:* Equipment that uses high, powerful frequencies can block and/or distort signals. Hospital imaging rooms, for example, can cause problems for other devices, especially the computer controlled devices found in security systems. Machine noise as well as the proximity of high voltage lines also disrupts RF.

- *Line-of-Sight Considerations:* Infrared wireless transmission is not bothered by electromagnetic interference, but does require line-of-sight transmission. Determine what could unintentionally block the transmission area.

- *Public Safety:* Check whether your RF devices interfere with PaceMakers or other health-related devices.

- *Distance:* Determine whether the RF device can cover the territory required and whether repeaters are necessary. Make sure your transmission area is well within the limits, with room to spare. "Just fits" usually don't.

- *Antennas:* Determine the size and types of antennas required and whether your premises can accommodate them. If antennas must be outside, determine whether local laws control their placement. If the antennas are outside your property line, check whether other property owners will allow you to erect them. An unwilling neighbor can destroy the best wireless system design.

- *Power Requirements:* All signals require power and power requires cabling or batteries. Make sure that enough power is available for the time during which the device will be used. RF frequencies that require more than one Watt also require licensing and other regulations. This may restrict the freedom of use for which the device was intended.

Analog-to-Digital

When a computer receives information from an analog signal, such as a standard telephone signal, it must translate that information into binary code. (See the *Index* for further information.)

Each point on an analog chart corresponds to a number and a moment in time. These unique numbers exist on a *table* (a chart of numbers) that the computer uses to translate analog numbers into binary numbers (or vice versa).

Analog-to-digital translations are made by a special computer chip that is embedded in a data acquisition circuit board. A similar chip is found in a *modem* (an acronym meaning "<u>mo</u>dulate - <u>dem</u>odulate") that connects computers through analog telephone lines.

Although the telephone system trunk lines are controlled digitally, *line card circuits*, which are the links between the general-use lines and specific customers, are mostly analog.

Modems are only needed when computers communicate through analog systems to other computers. Modems are not needed for digital transmissions, such as between devices in a computer network or through special digital telephone services.

Digital telephone networks do not require modems for data transmission, but *do* require a *codec* (an acronym meaning "<u>co</u>der - <u>dec</u>oder") for voice transmission. This works differently than a modem because it translates voice tones into digital information and vice versa.

Digital Transmission

Transmission speeds between computers are rated by the number of bits they can send per second, or *bps* for short. This rate of transmission is sometimes called a *bandwidth*.

As of 1996, modems can transmit data over standard telephone lines at the following bandwidths:

> 2,400 bps
>
> 9,600 bps
>
> 14,400 bps
>
> 28,800 bps - based on a protocol (*standard*) called V.34

As of this writing, a chip supporting a 57,600 bps transmission rate is being developed for use in 1997.

The problem with transmitting over telephone lines is getting those wires to carry data at consistent high speeds. This transmission is hampered by:

- Transmission technology,
- The quality of the trunk line switching technology and
- The volume of calls on the system, causing trunk line rerouting and/or information buffering.

When there is a heavy volume of calls, pauses occur in data transfer while waiting for trunk lines to open. When this happens, data is temporarily *buffered* (saved) along the route before it can move on.

One solution to bandwidth problems is *multiplexing* technology. This increases the "vocabulary" of digital information, thus providing more data in less time. Today, multiplexing technology allows a mixture of information — data, video, fax and/or speech — to be sent in one transmission over old-fashioned phone wire. It does not, however, necessarily speed things up.

Multiplexing can be used to transmit a variety of digital video signals in one data stream by patching these signals together. Transmission of one image, unfortunately, must be stopped before another can be sent. Needless to say, images are lost in the process.

Another solution is to eliminate modems.

In the early 1980s, a group of telephone carriers banded together to develop the Integrated Service Digital Network (ISDN) to speed the flow of data over standard phone lines. The objective was to bring the greatest improvement for the least possible cost.

ISDN technology improves the data transmission properties of standard copper wire telephone cable by making use of multiplexing technology, in addition to providing pure digital connections.

ISDN splits the services of common two-wire phone cable into three channels. Two channels provide 64 Kbps (64,000 bits per second) data transmission each and can be combined to provide 128 Kbps (128,000 bits per second). The third channel (D-channel) is used for call setup and signalling. It can also be used for voice through the use of a codec, mentioned earlier.

Higher speeds can be achieved when fiber optic cable is used; the actual transmission speed, however, is only as fast as the transmission, switching and receiving equipment will allow.

As of 1995, the highest transmission speed a modem could achieve using analog telephone lines was 28.8 Kbps (28,800 bits per second). Unfortunately, this speed cannot be consistently maintained because of data compression problems. Data compression works by reducing stretches of repeated information, such as "white space" through more efficient codes.

A wireless digital data and voice transmission system, pioneered by Metricom, located in Los Gatos, CA, claims to transmit data at speeds of up to 77 Kbps (77,000 bits per second), with a guaranteed throughput rate of at least 38.4 Kbps (38,400 bits per second). This is significantly faster than can be provided by a modem and is far less costly to implement nationwide than ISDN.

The problem faced by ISDN is that although the vast majority of trunk lines are digitized, the lines leading to individual residences and businesses are not. Converting these connections requires abandoning all existing telephones, installing new line cards at every point and investing in new digital receivers for everyone at a cost of trillions.

The implication of Metricom's success, then, is that wireless technology is challenging wired systems. Wireless transmission, of course, requires antennas and some communities do not want antennas erected. Consequently, in 1995, the United States Government began debates on antenna placement and licensing to safeguard that interstate data transmission would continue to improve — especially important to national security!

Dedicated computer networks, however, transmit data at far higher rates than is done over public carriers. These rates can range from 1 Mbps to 100+ Mbps, depending on the networking system. Like transmission over public carriers, however, speed degrades when many people are using the network at the same time. Like telephones, if a receiving computer is busy, data cannot be transmitted at all.

Data Transmission Carriers:

Data networks, such as between security substations, currently make use of a number of wired and wireless transmission carriers, depending on need and budget. These include:

Computer-to-Computer (networking) - dedicated systems which include cable and wireless modes of communications. These transmit data at the fastest rates. Acronyms include LAN (local area network), WAN (wide area network or building-to-building), and MAN (metropolitan area network or communicating throughout a geographic area).

Public Switched Telephone Network (PSTN) - standard phone lines with a maximum bandwidth of 28.8 Kbps based on V.34 protocol.

Integrated Services Digital Network (ISDN) - standard phone lines controlled exclusively by digital switching equipment which provides two 64 Kbps digital channels, plus one channel for signal connections and voice transmission. A bandwidth of up to 128 Kbps can be achieved by combining the first two channels. *Note:* A codec is required on this digital network in order to send voice.

Digital Data Service (DDS) - a special leased line (dedicated line) that offers either 56 Kbps or 64 Kbps bandwidths.

Switched 56 Network (SW56) - a dial-up service that offers a 56 Kbps bandwidth. A special code number must be dialed to connect into the network before the destination number can be entered. This lets users pay only for the time they use without requiring a full-time leased line.

Digital Cellular Service - wireless data transmitting service with a 19.2 Kbps bandwidth.

Digital Spread Spectrum Service - wireless voice and data transmitting service based on spread spectrum technology with a bandwidth up to 77 Kbps and guaranteed throughput of 38.4 Kbps.

Satellite - wireless transmission that carries data long distances at speeds comparable to those of DDS or ISDN lines. This is not, however, the best way to transmit data as pauses in transmission can significantly increase transmission time, thereby reducing transmission speeds or destroying it altogether.

Improvements in transmission speeds and switching technologies are being made daily. The race is on.

Estimated Data Transmission

The chart on the next page, contributed by Herb Guenther of WebZone Communications, estimates the times required to send various amounts of data using current data transmission technology.

Assumptions made in developing the chart:

All protocols listed on the chart are for serial communications. Most data is sent in 8-bit bytes, so data sizes are listed in bytes. To each byte is added one start bit and one stop bit. We assume that parity bits are not used. The setting of N81 for no parity, 8 bits per byte and 1 stop bit is the most common and the assumed parameter.

We assumed no data loss, that all information sent is received, and no retransmission of data is needed. This is not usually the case, a good connection needs very little retransmission, while a poor link can double the transmission times listed.

We assume that the negotiate speed of the link is the maximum listed and that both ends have the same speed equipment, otherwise the lower of the two ends will govern the connected speed.

We recommend that files be compressed before sending. This reduces file size and therefore, reduces transmission time.

To use the chart, pinpoint the technology you will use in the far left column and the size of the file to be transmitted in the appropriate column to the right.

The estimated time can be frustrated by such variables as busy lines and other transmission problems in the system. Keep in mind that the more popular the system, the heavier the traffic. This causes backups or the complete unavailability of the line required.

All transmission speed depends on sending and receiving at the same rate. High speed modems connected to lower speed modems only send data at the lower speed.

Importance of Communication Speeds

The chart on the next page will allow you to calculate rule-of-thumb values to determine potential bottlenecks and delays in your communication system.

Monitoring the actual transmission speeds in your system will allow you to spot problems and fix them before the system completely fails.

Estimated Data Transmission Times

bps=bits per second   mbps=mega bits per second   gbps=giga bits per second

	Transmission Speed	Volume of Data Transmitted in Bytes				
		100K 1,024,000	1 Meg 10,240,000	10 Meg 102,400,000	100 Meg 1,024,000,000	1 Gig 10,240,000,000
Analog Communications - using modem	300 bps	56.9 min.	9.5 hrs.	4.0 days	39.5 days	395 days
	1200 bps	14.2 min.	2.4 hrs.	23.7 hrs.	9.9 days	98.8 days
	2400 bps	7.1 min.	1.2 hrs.	11.9 hrs.	4.9 days	49.4 days
	9600 bps	1.8 min.	17.8 min.	3.0 hrs.	29.6 hrs.	12.4 days
	14,400 bps	1.2 min.	11.9 min.	2.0 hrs.	19.8 hrs.	8.2 days
	28,800 bps	35.6 sec.	5.9 min.	59.3 min.	9.9 hrs.	4.1 days
	57,600 bps	17.8 sec.	3.0 min	29.6 min.	4.9 hrs.	2.1 days
ISDN (digital)	64,000 bps	16.0 sec.	2.7 min	26.7 min.	4.4 hrs.	1.9 days
	128,000 bps	8.0 sec.	1.3 min.	13.3 min.	2.2 hrs.	22.2 hrs.
Ethernet local area network	10 Base T 10 mbps	.1 sec.	1.0 sec.	10.2 sec.	1.7 min.	17.1 min.
	100 Base T 100 mbps	.01 sec.	.1 sec.	1.0 sec.	10.2 sec.	1.7 min.
Point-to-Point wide area network	56,000 bps	18.3 sec.	3.1 min.	30.5 min.	5.1 hrs.	2.1 days
	T1 line 1.5 mbps	.7 sec.	6.8 sec.	1.1 min.	11.4 min.	1.9 hrs.
	T3 line 10 gbps	.0001 sec.	.001 sec.	.01 sec.	.1 sec.	1.0 sec.

Transmission Considerations

Wireless systems are used when the cost of wiring is prohibitive or too complex to maintain. Whether the need is present or not, however, today almost any device that sends information can be designed to do so wirelessly.

Unlike their cabled cousins, however, radio signals can be more easily contaminated by *noise* (lightning, nearby power cables, machinery) and consequently, cannot provide fail-safe performance. In addition, because frequencies are generally available "in the air," many receivers can capture them, keeping spies well employed. This is a particular problem with cordless room monitors, phones and cellular services, to name a few.

The military has the highest need for confidentiality and so has developed many jam-proof RF systems, which are not available to the general public. As they develop better systems, however, their older technology is released to the general market.

Spread spectrum, for example, was developed during World War II (1940s) and was declassified in the mid-1980s. It provides jam-proof security by broadcasting a single message over multiple wavelengths. Each part, then, is sent over different frequencies, making it impossible for a spy to patch together the broadcast without a one-of-a-kind receiver.

Access control devices using spread spectrum operate in the 902-928 MHz range and can transmit to distances of 3,500 feet or more with special antennas. The FCC does not require licensing as long as its maximum transmission power is restricted to 1 Watt. Computer networks and digital services using this technology broadcast at higher bandwidths.

Spread spectrum is now playing an important part in the communications industry because of its high level of security. Its technology is used in data transmission services such as Metrocom, mentioned earlier in this chapter, wireless computer networks, wireless phones and, of course, wireless access control.

Problems with Signal Carriers and Receivers:
As stated earlier, connections (wired or wireless) require a transmitter, an information carrier and a receiver.

General broadcasting sends its signals as a one-to-many. This means that one broadcast will be received by anyone who is tuned on an appropriate receiver.

Private communications, however, are usually transmitted as one-to-one, or one-to-a-select-few. This means that when all the carrier lines are used, no new transmissions can be made. Likewise, if the target receiver is in use, no transmissions can be made, no matter how many lines are available.

Phone lines and cellular services in large communities jam during peak hours. This even happens in small communities where the trunk lines haven't kept up with the growing population, or there is a sudden surge of visitors, as happens during the summer with tourists.

In addition, computers are being called upon to monitor more and more areas. If the system isn't distributed, or too many remote computers are trying to send alarms to the same host computer, jams occur.

Security professionals need to calculate the effect of receiver availability and the potential of line-jams when designing their EAC communications network. The fastest data transmission system in the world is only as good as its ability to make a connection.

Transmission Consideration Summary:

- *Information Priority:* Determine what information absolutely must be transmitted at all costs and what information can be sent at any time.

- *Carrier Lines:* Determine how many lines you need and what might be the outcome of having too few lines or lines that can experience jamming.

- *Receivers:* Determine how many receivers you need. Does one computer have enough I/O to receive data simultaneously from numerous sources? What other options are available?

- *Actual Transmission Time:* Determine the amount of transmission time you need for each task and the data transfer speeds required to make those transmissions acceptable.

- *Confidentiality:* Determine what level of confidentiality is required.

- *Security:* Determine how sabotage could affect your system.

- *Redundancy:* Determine what backup or alternative methods are needed if your system is somehow breached by line-jamming, electrical interference, or by a spy.

Chapter Review - Communications

Use cameras and video tape recordings to . . .

document your connection inventory and to train new people as to how your communications system is set up.

Every change to your system requires documentation because . . .

the technical information has to be efficiently passed on to new personnel and maintenance people.

Regularly scheduled wiring audits are performed to check whether the:

- wiring is healthy (not corroded),
- connections are solid,
- connections meet code and
- system has been exposed to tampering.

Some characteristics of cable housings include:

1. Color coding.

2. Flexibility to respond to atmospheric conditions (heat/cold).

3. Strength to withstand pulling.

4. Waterproof capability for outside installations.

Four types of cable commonly used in communications are:

- Unshielded Twisted Pair (UTP)
- Shielded Twisted Pair (STP)
- Coaxial Cable
- Fiber Optics Cable

Conduit categories are:

- Residential (CM-X) - lowest grade
- General Purpose (CM or CM-G) - commercial grade
- Riser-rated (CM-4) - runs up or down walls
- Plenum (CM-P) - runs in ducts and vertically above the ceiling

Troubleshooting tips are important because . . .

> they can help you plan your installation as well as maintain it. (Check the *Index* under "tips" or "troubleshooting.")

Wireless connections can provide . . .

> device portability (no dependence on wires), accessibility (put things where you want them) and creativity (explore new options).

The communications bandwidth assignments, as provided by a chart in this chapter, show you . . .

> what bandwidths are accessed by the radio frequency devices you use.

Factors that impact radio transmission are:

1. Government regulations and licensing.

2. Surrounding building materials.

3. Surrounding natural environment.

4. Weather.

5. Surrounding equipment.

6. Line-of-site clearance.

7. Public safety.

8. Distance.

9. Antennas.

10. Power requirements.

While it is useful to get the most powerful modem possible, the actual speed of data transmission depends on . . .

> waiting for direct telephone line circuits to open during a heavy volume of calls.

Multiplexing technology does not speed up communications, but it does let you . . .

> send a greater variety of signals, such as data, video, fax and/or speech, in a single transmission over standard telephone lines.

Consulting a list of Data Transmission Carriers, such as found in this chapter, lets you . . .

> pinpoint the type of service you need before making general inquiries to various carriers.

The chart "Data Transmission Time at Various Speeds," found in this chapter, is a guide to help you calculate the . . .

> time required to send an approximate volume of data at a specific transmission speed.

QUESTIONS

1. Why are facility maps helpful to have?

2. How does unshielded twisted pair wire work to cancel electrical signal interference?

3. Why do installers have to worry about conduit size and cable pulling factors?

4. If a modem is used between a digital computer and an analog phone line, what type of device is used between a digital computer and a special digital phone line?

5. If you know how much data is flowing between two points, how can the chart called "Data Transmission Time at Various Speeds" help you decide what components will best make up your network?

Chapter 8

System Design
A Technical Design Perspective

By Warren Simonsen

EAC System Basics

Up to this point, this book discussed many of the components used to create an electronic access control (EAC) system. This chapter describes the issues involved in combining these elements into a system.

A *system* is a collection of components that functions together for a single purpose. Obviously, the single purpose of an EAC system is to control physical access to a building or other facility.

This single purpose becomes less obvious when EAC systems are integrated with other applications, such as those that monitor fire detection devices, regulate air flow and temperature, control lighting, monitor asset location, and provide tracking of attendance and hours.

Unfortunately, the term "integrated" has been abused almost to the point of meaninglessness, but for the purpose of this book, *integrated* refers to any system that is controlled by a single software product.

This chapter discusses the design of EAC systems, including the integration of access control, intrusion detection, and CCTV control. This does not mean that systems with wider ranges of integration are not common, only that those systems are outside the scope of this book.

System Design Goals

From a technical standpoint, all properly designed EAC systems have at least two things in common:

1. Systems are sized to meet the needs of the intended function, and

2. Systems have safeguards against failure.

System Size

A properly sized system meets the needs of the intended function. It allows for future growth, but does not saddle the system with unneeded equipment.

Sizing choices are made for now *and* in the future. They include the number of:

- Doors to be controlled
- People who will use those doors
- Sensors to be monitored

As systems get larger, the technical choices become less obvious. The system designer must also take into account the:

- Amount of data that is likely to be transferred between devices.
- Compatibility between devices made by different manufacturers.
- Installation issues, such as power and electrical grounding.

This chapter discusses the technical choices in designing a properly sized system. The next chapter discusses the system design from the functional perspective.

System Safeguards

Systems are designed to work, not fail. Unfortunately, in the real world, failure occurs too often and at times and places that can be highly inconvenient. Therefore every system must be designed to fail gracefully. This means that as components fail, the system should continue to provide as much functionality as possible.

Planning for graceful failure means that the designer must review each component of the system and determine the resulting operation of the system should that component fail. If the resulting operation is unacceptable, redundant or alternate methods must be added.

The good news is that today's distributed intelligence systems are designed to be redundant. They generally allow for multiple component failures before facility security is compromised.

System Components

A typical EAC system is comprised of the following components:

Control Panels, where the decisions of granting or denying access and declaration of alarms happens in real-time. Each panel controls a specific number of doors, usually from one to eight, and monitor a specific number of sensors.

Output Devices, such as locks, buzzers and lights. These are typically activated by the Control Panel based on activity in the rest of the system.

Input Devices, such as intrusion detection devices, door status monitors and credential readers. These are the eyes and ears that allow the system to determine the current state of the system and any requests for changes to the system such as a request to open a door.

Supervisory Computers in today's systems provide a central means for security personnel to define and monitor the operation of the system. These computers can be personal computers running a wide variety of operating systems, UNIX workstations, or even mini-computers.

Supervisory computers let operators easily manage large and complex systems by prioritizing events and supplying directions for the operators. The use of computers makes it easier to integrate the creation of credentials and the control of CCTV systems.

These computers often have the following external devices connected to them: printers, scanners, signature capture pads, video capture devices and backup storage units.

The **Software** that runs on the supervisory computer is one of the most crucial items in the system because it defines the operator's use of the system. The software determines the way operators monitor and respond to events. The software generates reports management uses to track security and is the means by which security personnel define the system's operation.

Communications Devices, such as port expanders, modems, repeaters, bridges, gateways, and routers, are used to connect the various components of the system to allow the exchange of data. The types of networks chosen, the cabling, and connectors are important to a smooth running system that is easy to maintain.

The issue of **compatibility between components is critical.** Always check the manufacturer's specifications for system compatibility. The hopeful "it should" work is usually not good enough to insure that it will work.

Computer Networks

The EAC computer network is the communication system between the computers used to supervise the EAC system. This may or may not be part of a general purpose computer network.

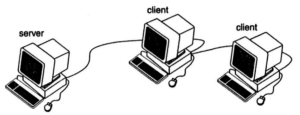

A general purpose network connects computers and their peripherals, such as printers and scanners.

Making the EAC computer system part of a general purpose network can save cabling, equipment costs, and leverage the efforts of people already supporting the network. On the other hand, the EAC network is sensitive to information delays. As we discuss later in this chapter, you don't want to find out about an alarm minutes or hours after it occurs. Also, combining EAC with a general purpose network opens the security system to attack from any user on the general network.

Many EAC systems use one supervisory computer and would not require a computer network; however, the trend toward computer networks is growing. If you have more than one supervisory computer, a network is essential.

Client / Server

If you spend any time around computer networks, you will hear the terms "client" and "server."

> The *client* is the computer requesting a service or data.
> The *server* is the computer supplying it.

Many networks have one or more computers designated as "the server." The primary purpose of a server is to function as a central depository of files and/ or services. These central services may include supporting printers, providing a connection to the Internet, sending and receiving FAX transmissions, or acting as the post office for electronic mail.

The other computers on the network, or clients, are generally regarded as personal workstations and are used for everyday tasks by the people accessing the network.

To make the issue more complex, a server can also function as a personal workstation, but that is typically avoided because the server's processing time is needed to quickly respond to the requests of the network.

EAC Adds Control Panels to a Network

As we mentioned, a computer network can have more than one server. An EAC system will typically have a computer designated as the EAC server where the data base is kept. The EAC server may or may not be the same computer as the network server.

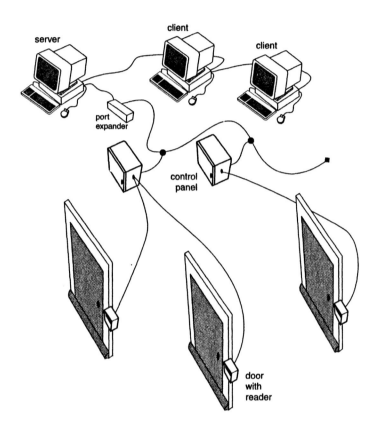

It is perfectly acceptable, even recommended, that the EAC server actually be one of the machines that functions as a client for the general network. This is recommended because the EAC system can have large numbers of transactions that will consume the processing time of the EAC server. If this machine is also functioning as the general network server, all users will experience slow response times. When network users come looking for who is slowing down the system, it is not a pretty sight.

Network Protocols

There are many different *protocols* (rules) followed by the communications software for computer networks. Two of these protocols are so common and referred to so often that they deserve mention here:

- **TCP/IP** is the protocol used by the Internet and, as a result, by many other applications as well.

- **IPX/SPX** is the protocol used by Novell networks. Novell has been a longtime leader in the installation of networked systems and many corporate systems depend on this protocol.

These protocols *are not* compatible in the sense that software designed to use one of the protocols will not be able to make use of the other protocol.

They *are* compatible in the sense that both protocols can exist on the same cable and computers can be set up to respond to both protocols. It is possible to send TCP/IP packets on your Novell network provided the sender and receiver have been set up to handle the TCP/IP protocol.

One possible problem with mixed protocols on one network is a device called a router. A router is a dedicated computer that monitors messages being sent from one leg of a network to another. It passes messages that are intended for a node on another leg and does not pass messages that are intended for nodes on the same leg as the sender.

Routers are very helpful in keeping the traffic on a network to a manageable level. The problem is that routers must know something about the protocol to be able to open a message and determine the destination. A router on a Novell network, for example, may not be able to process TCP/IP messages and therefore not pass them to the next network segment.

Expert Help

You may need help from a computer consultant or your Information Systems department to install a computer network. This is particularly true if you are going to use an existing computer network to also support the EAC system.

For more information on computer networks, we recommend the following books:

How Networks Work, by Frank J. Derfler, Jr and Les Freed, published by Ziff Davis Press

> This is an easy-to-understand book that is highly illustrated. It lets you see the relationship between components and introduces technical concepts with a minimum of words.

Guide to Connectivity, by Frank J. Derfler, Jr., published by *PC Magazine.*

> This provides a detailed, broad view of networks.

The Essential Client/Server Survival Guide, by Robert Orfali, Dan Harkey, and Jeri Edwards, published by Wiley Computer Publishing.

> This covers networks, protocols, and the ways in which they are applied.

Control Networks

The control panels in an EAC system can be thought of as special purpose computers. As such, many of the same principles used in computer networks also apply to networks connecting the control panels (control networks). Since the control panels are highly specialized, the protocols tend to be proprietary, or specific to that manufacturer.

Computers from different manufacturers communicate over most computer networks. In a *proprietary network*, such as those used by EAC panels, only equipment built by the manufacturer of the network will function properly.

In recent years standards for control network protocols have been proposed which will allow equipment from different manufacturers to communicate over the same network. These universal control network protocols are slowly achieving acceptance.

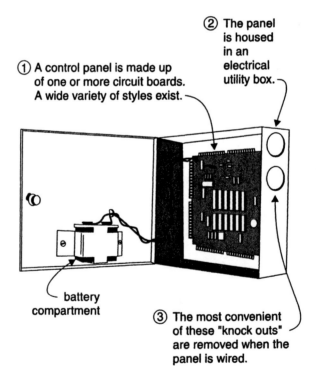

① A control panel is made up of one or more circuit boards. A wide variety of styles exist.

② The panel is housed in an electrical utility box.

battery compartment

③ The most convenient of these "knock outs" are removed when the panel is wired.

When we talk about connecting control panels to a network, we refer to each panel as a *node*. The word "node" is a generic term for any device connected to a network.

In order to communicate, each node has a *transmitter*, used to send information, and a *receiver*, used to receive information.

There are typically two methods for connecting a *node* in control networks: *multi-drop* and *daisy-chain*.

In a multi-drop system, the transmitter and receiver of each node are connected over a common conductor with the transmitter and receiver of every other node.

The daisy-chain, or "loop," method has the transmitter of the first node connected to the receiver of the second node. The transmitter of the second node is then connected to the receiver of the third node and so on until the transmitter of the last node is connected back to the receiver of the first node completing the "loop."

Multi-Drop

The multi-drop method is popular because it is easy to add or remove a node from the network.

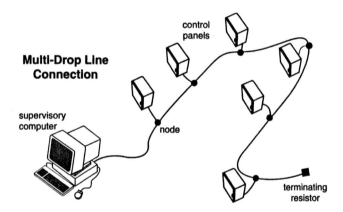

Because all the transmitters and receivers share a common conductor, the multi-drop method requires a means of insuring that only one transmitter is operating at a time. This is handled by the network communications software and is invisible to the user, but knowing that it exists helps to understand some of the requirements and troubleshooting techniques for this method.

Generally a multi-drop scheme requires a resistor, called the terminating resistor, at the end of the network cable.

As the network gets larger, repeaters may be needed to rebroadcast and strengthen the signal. A popular multi-drop communication standard, called RS-485, calls for a repeater for every 4,000 feet of network cable length or for every 32 nodes. Consult the manufacturer's specifications for your control network because these limits can depend on many different factors.

When a node in a multi-drop network fails, only communications to that node is lost. The network will continue to function normally minus that node.

When the cable is cut, all nodes downstream, from the cut to the terminating resistor, are lost along with the terminating resistor. The loss of the terminating resistor degrades the performance of the network and may result in slow or lost communications. Usually some communication remains with the nodes upstream of the cut.

Daisy-Chain

With the daisy-chain method each transmitter can transmit whenever it pleases because the information it sends is picked up by only one receiver.

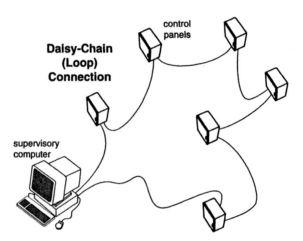

Daisy-Chain (Loop) Connection

control panels

supervisory computer

In a multi-drop system, a repeater is required to strengthen the signal over long distances. Since each transmission between nodes in the daisy-chain method is a rebroadcast of the signal, there is much less need for a repeater.

Even with each node rebroadcasting the signal there are still limitations on the number of nodes and the distance between nodes. Again, the manufacturer's specifications are important because of the number of factors that determine the network limits.

In a daisy-chain network the effect of losing a node is the same as cutting the cable.

Depending on the network protocol used, the *least* effect of a cable cut or node failure is loss of communications *to* every node downstream and loss of communications *from* every node upstream.

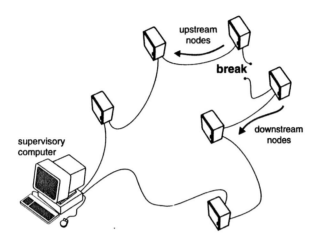

Control Network Safeguards

For both the multi-drop and the daisy-chain methods, it is possible to add redundant paths to keep the network fully functional in the case of a cable cut, but this is more expensive.

The benefit of distributed systems is that even with the loss of communications, intelligent control panels continue to provide access control. They also store a record of the events monitored by the panel in a memory buffer until communications are restored.

ASK THESE QUESTIONS:

When designing the system, ask the supplier about the operation of the system if a node fails or if the cable is cut. Make sure that these events are immediately detected by the system and reported to the operator for corrective action.

Ask about the panel's memory buffer size in terms of event records stored. The size of the panel's memory and an estimate of the number of events that occur while the panel is disconnected should give you some idea how long communications can be disrupted without losing any records of the events that occurred.

Inter-Network Connections

The connection between the Control Network and the supervisory Computer Network can be accomplished in many different ways. Due to the fact that most computers have serial communication ports, the connection is generally made through those.

Serial Connections

Serial ports, also called *com ports* ("com" stands for communications) are the typical means of connecting various computer based devices.

Serial ports on personal computers, workstations, and most mini-computers use the RS-232 standard for serial communications. This standard has been around for a long time and many devices have been built to use it.

The RS-232 standard only defines the electrical and mechanical characteristics of the connection. It does not insure that two devices, each using an RS-232 connection, can actually communicate properly. Successful communication is determined by the software in the devices at each end of the connection.

EAC systems typically have many devices that require a serial connection while personal computers have a limited number of serial ports. Since each serial port can support only one device, port expansion devices for the PC have been developed. These devices consist of a card that plugs into the PC's main circuit board. This card is then connected to a separate utility box that contains the additional serial ports.

Another standard for serial communications is RS-485. It is possible to connect a Control Network, for example, using RS-485 directly to the computer, but most computers do not come equipped with hardware that supports the RS-485 standard. To connect devices using RS-485 to a personal computer, a compatible interface board will need to be installed. You must also make sure that the software on the computer can support the interface board.

Modems

It is not always possible to directly connect the Control Network to the Computer Network.

Multiple Control Networks as well as long distances sometimes dictate the use of "temporary" connections through dial-up telephone lines or the use of long-distance communication systems leased from a third party.

> *Example:* Global Manufacturing, Inc. is located in a large city and all EAC functions are monitored by a central security computer. This com-

pany also has a storage facility in another country, where EAC traffic is light.

The control panels at the remote facility are hooked up to a modem. Periodically the security computer uses its own modem to connect with the modem at the remote facility. Once the connection is made, the security computer polls (contacts) each control panel at that site for a report of activity.

In case of an emergency, the Control Network at the remote site makes use of the modem to dial the security computer and report the situation.

A modem is a special circuit card that controls the flow of information over commercial telephone lines. These lines were designed to transfer the analog signals that are produced by telephones carrying voice messages. The modem converts the digital signal from the computer into an analog signal for the telephone line. The modem on the other end converts the analog signal back into a digital form for use by the receiving device.

There is also a device known as a codec that transmits and receives digital signals over the telephone lines. It is possible in many areas to request an ISDN line which is a telephone line designed to carry digital signals, but again, because it is not yet common, the cost is significantly higher.

Other Connections

There are many other communications standards other than RS-232 and RS-485. These other standards require specific interface boards and supporting software to create a functional system. The supplier of the Control Network will provide the interface hardware and software for the supervisory computer.

It is possible for the Computer Network and the Control Network to be the same network. In this case the control panels and other devices on the Control Network are connected directly to the Computer Network and only the software determines whether messages are between the supervisory computers, between the control panels, or between the control panels and the supervisory computers.

This type of connection for control panels is more expensive and is generally only cost effective if the Computer Network already exists everywhere control panels are needed and the Computer Network is not heavily used.

Network Recommendations

We highly recommend that the Control Network be separate from the Computer Network to insure that the report of a significant security event is not delayed by heavy network traffic and to make it more difficult for computer hackers to compromise the security system.

For these same reasons we also recommend the Computer Network that connects the supervisory computers for the security system be separate from the corporate or enterprise-wide network. If these networks are combined careful attention must be paid to traffic volume and security concerns.

Connections in General

The quality of cables and their connections are as important as the computers and EAC devices in the system. The following items are very important when designing any electronic system:

- Cables should be well labeled and installed with care. Make sure that your installers know and follow the appropriate electrical codes. Poorly installed cable can create electrical short circuits or breaks in the wire that are very difficult to diagnose and locate.

- Unless you are using fiber-optic cables take care to avoid sources of electromagnetic interference (EMI). Typical sources of this electrical noise are motors, fluorescent lighting ballast, and transformers. Avoid running cables carrying data signals in the same conduit as those conducting power. If data cables must cross power cables, do so at right angles to minimize the interference.

- **Pay attention to specifications.** A manufacturer's specification sheet indicates the limits to which a piece of equipment has been tested. Failure to follow specifications may appear to work for a while, but chances are conditions will eventually occur that will cause the system to fail. Finding problems in a system that are caused by failure to follow specifications is difficult and usually very expensive.

Review Chapter 7 - *Communications* for further details about cable types and other information about correct installation.

Software

The supervision of every system has three components. These components are defined in the software so they are not always easy to separate. Understanding that every system has these three components, prepares you to understand and compare different implementations. These three components are:

the *user interface*,

the *data base*, and

communications.

All three of these components can exist on the same computer, but in very large applications each of these components might require a separate computer.

User Interface

The user interface determines how software is seen and used by people. This includes everything from the way information is presented in reports and on the screen, to the selection of which information and options are presented to the user and when.

Many software packages today include a Graphical User Interface (GUI, pronounced "gooey") which can improve the ease of use. A GUI also requires a more powerful computer and better monitor than a standard text-based interface.

Don't let the color and flash of fancy graphics determine your choice of a software package. Pay close attention to the information conveyed and compare it to the way you will use the system.

Elaborate interfaces that are confusing detract from the usefulness of the software. Colors or animation that are pretty, but don't convey information will eventually annoy and distract the operator from the information needed to use the system.

When evaluating software, spend some time with the user interface looking for clarity. Information that is used regularly should require a minimum number of keystrokes or mouse clicks to access. This is where an integrated system is most helpful because the user does not need to leave one software package and start another to accomplish a task.

Also evaluate typical operator actions in an EAC system which include:

Alarm monitoring,

System control,

Report generation, and

Operator security levels.

Systems with higher levels of integration may also include the creation of credentials or badges and the automatic control of CCTV cameras and other equipment within the user interface.

Alarm Monitoring

Even if there are no general purpose intrusion detection devices in the system, almost every EAC system monitors the doors for forced entry.

The alarm monitoring screen should clearly indicate to the operator which inputs are in the alarm state both visually and by a sound that is not easily ignored. Typically this is how it works:

Each alarm requires the operator to acknowledge the event by pressing a key or clicking a mouse on an ACK (acknowledge) icon. Many times the operator receives instructions on what to do when this alarm occurs and may be required to enter a description of the event and the operator's response.

Acknowledging the alarm should change its display and make it less obvious than alarms that have not been acknowledged.

After the operator has acknowledged the alarm and handled its cause, the alarm may be cleared from the display, but is stored permanently in a history file.

In addition to "alarm" and "normal," *supervised inputs* may indicate a "trouble" state. A supervised input is one that is electrically monitored to prevent tampering. If an attempt is made to bypass a supervised input or if something happens to damage the circuit, the display will indicate that point in a trouble state.

Operator Control

Most EAC systems operate automatically; reading credentials and making decisions about granting or denying access. From time to time human intervention is required, such as when someone forgets their access credentials or an event occurs that was not covered in the system's automatic programming.

In these cases it is important that the operator be able to easily and smoothly control the system in a manual fashion. Consequently, the operator must be able to unlock and/or open doors, shunt alarms, and perhaps activate other devices such as cameras or lights.

Shunting an alarm is a term used to indicate that the alarm input is to be ignored. For example, the open door alarm is automatically "shunted" when access is granted to allow the door to be opened without falsely signaling an alarm.

Other control terms used in these systems are energize, de-energize, and pulse.

If a device is *energized*, it will stay that way until *de-energized*. Whether "energizing" locks or unlocks the door is defined by each system.

If a device is *pulsed* it will be energized for some defined time period and then automatically be de-energized. Pulsing is convenient if you wish to allow one person through a door. Pulsing the lock allows the door to be unlocked for that person, but does not require the operator to remember to re-lock the door after the person enters because the output is returned to the original state automatically.

Access Definition

Access control is managed by three concepts:
- Access Level
- Time Zone
- Access Area

EAC systems determine when an individual can enter an area by assigning an *access level*. An access level is a combination of a time zone and one or more access areas. Access areas group the use of specific readers.

Time is managed in an EAC system through the creation of *time zones*. Time zones are periods of time given a label so they can be referred to in a systematic fashion.

Example:
The First-Shift time zone may refer to the time periods between 7:00 am and 5:00 pm on Monday through Friday. Time zones must take into account the time, the day of the week, and holidays.

Chapter 8

The controlled areas in a facility are referred to as *access areas*. Each access area is defined by the credential readers assigned to it, therefore, access areas and physical rooms may not necessarily coincide or they may overlap.

If the computer room supervisor's office is inside the computer room and both rooms are controlled by the EAC system, the definition of the access area is determined by the assignment of the specific readers.

Example 1
If the reader for the computer room and the supervisor's office are both included in one access area, a person in the supervisor's office would be considered inside the computer room access area.

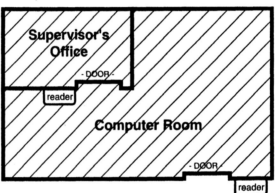

Example 1 - One access area.

Example 2
If an additional access area is created for only the supervisor's office, the person in that office would be considered to be in both access areas at the same time.

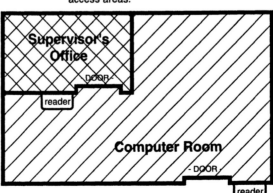

Example 2 - This room-within-a-room creates overlapping access areas.

Example 3

If the reader for the supervisor's office is then removed from the definition of the computer room access area, the person in that office is considered to be in the office access area, but not in the computer room access area.

Example 3 - Two unrelated access areas.

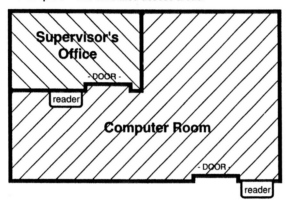

During all this, the rooms and the person did not move. The definitions depend on how the operators think of the facility.

Entrance to each access area is determined by an *access level.*

Example:

George may be assigned the access level of Supervisor which entitles him to enter any door on weekdays except the Executive Washroom. Sam, on the other hand, is assigned the access level of janitor which allows him access to every door including the Executive Washroom but only between the hours of 8:00 pm and midnight.

Access levels allow the assignment of common time and door combinations to multiple people without having to individually define that combination for each person. Access levels also allow management to make fast changes for a large group of people.

Example:

If the starting time for first shift changes, the access level can be modified and everyone affected is updated without having to open each person's record individually.

Report Generation

When evaluating a system, determine the kinds of reports you are going to need.

Creating a history report of the events that occurred is one of the reasons many people install EAC in the first place.

Creating a report of the people who hold or held credentials is also important because it lets you keep track of system users.

Equally important is a report of the configuration, or description of the system's operating characteristics. These operating characteristics are determined when the system is installed.

The number of reports and the way in which they can be organized is limited only by your imagination. The important factor in selecting an EAC system, then, is whether the system provides, or can provide, the reports you need.

Operator Security Levels

Not all operators are created equal. It is common to have a group of operators who are required to monitor alarms and perform some manual control of the system, but are not allowed to change the system configuration or add new credential holders.

When checking new software, determine whether the system allows you to restrict operator access to various parts of the system in the manner that suits your needs.

Emergency

The determination of what constitutes an emergency and what steps are required to respond must be determined prior to selecting an EAC system. Your EAC system should allow you to anticipate a broad range of emergencies and provide a means to instruct operators what to do when they happen.

To help you evaluate this capability consider the following questions:

> Can your EAC system tell you who is in threatened areas when an emergency occurs?

> When an alarm occurs, are the appropriate steps for response clear to the operator?

Emergencies are unusual events and, because of this, the responses are not well practiced. The software must be clear and easy to understand so that mistakes are not made in the excitement of the moment.

Data Base

EAC systems have a lot of information that is appropriate for data bases. A data base is simply a table of information that can be accessed and sorted in multiple ways.

Data base software packages for PCs are available at any store selling software. These products are like toolkits; by themselves they can be used for EAC, but you have to figure out the data you need to store and create the tables and interface yourself.

EAC systems make use of the data base engine from these products to create an application that has this work already done. The *data base engine* is the core program of the commercial product without the user interface.

Data base software is fairly mature and most EAC systems use a commercially available data base engine. The EAC system benefits from the data base product's experience with handling data quickly and reliably.

Use of a commercially available data base engine might also make it easier to import or export data from the EAC system which is useful if data is to be shared with Human Resources or other data-intensive systems.

EAC-related data includes:

- Programming configurations for devices in the EAC system.

- Time zones and access level definitions.

- Access area definitions where the credential readers are assigned to specific areas to be controlled and tracked.

- Detailed information on the credential holders. Since the large majority of credentials are cards, this is typically referred to as the "card database." This will include personal information, the credential issued, the credential's standing (i.e., active, expired, or lost), and sometimes the photograph, signature, or other biometric information.

- Event logs that gather information on access transactions, device activity, and other system events.

- Communications interfaces such as the addresses of the control panels, the port they are connected to, and for temporary interfaces such as a modem the connection information (telephone number).

- System logs that track computer use, operator logins, and changes in configurations.

- If the EAC system integrates a CCTV system, it must keep track of camera locations, available monitors, and actions to be taken when specific events occur.

The data base portion of the EAC system determines how fast information can be accessed and the reliability of that data during operation and especially when the computer is unexpectedly turned off or loses power.

Data bases are especially susceptible to damage on power loss. It is a good practice to understand the likely effect a power loss will have on your data base and what steps should be taken if the data base is corrupted (damaged). Any reasonable data base will have automatic or manual methods of analyzing, isolating, and perhaps even correcting the damaged data.

Extremely Important
It is important to check and repair a data base as quickly as possible when you suspect corruption. Continuing to operate a damaged data base has the distinct possibility of further corrupting good data.

Communications

The communications portion of the software is the least visible but most crucial to the correct operation of the system. This is the traffic cop that determines which information flows first, where it goes, and even insures that the information was delivered correctly.

Earlier discussions about connections indicated that the software dictated the devices and communication standards that are supported. This is the part of the software that determines those options.

The product specification will indicate the number and types of devices supported and the standards that are followed.

Data is typically transmitted serially. This means that the bits of information that are transferred between devices flows one after the other down the transmission cable. Even local area networks transfer information serially, although at very high speeds.

Usually there can only be one message at a time on any one cable. To get a better picture of what is happening in the communication system let's use an analogy.

Example:
The communication cables are like train tracks. The messages being conveyed are the trains. The longer the message, the more cars on the train.

The various devices in the communication system, such as the control panel or the repeater, are the terminals where trains either unload their cargo (message) or are transferred to another track (cable) to continue the journey to another terminal (device).

Some of the tracks are relatively slow (thousands of bits per second) while others are faster (tens or hundreds of thousands bits per second). Some tracks, like those used in local area networks, are very high speed (tens or hundreds of million bits per second).

Each of the devices must track which train, or message, is allowed to travel which track (cable). The devices must insure that the high priority trains are allowed to travel first and that there are no collisions due to multiple trains being allowed on the track at the same time. A long train uses the track for a longer time making it unavailable to other trains.

Hopefully this makes it easier to understand that determining the overall speed of data transmission is a combination of:

- the transmission speed of the systems components,

- the length of the messages,

- and the routing ability of the devices that control the communications.

The speed of data transfer is crucial in EAC systems.

An EAC system is monitoring *real-time* events. This means that when an alarm occurs, the operator must be notified at that time, not some minutes or hours later.

> "Real-time" in an EAC system is determined in terms of human response time or essentially tenths of a second.

Obviously as alarms pile up rapidly on the display, the operator's response time for each event will slow down as time is spent analyzing and responding to each alarm. The software must continue to receive the events as they occur and not delay or lose event transactions.

A properly sized EAC system must be able to handle the worst-case scenario of every alarm occurring at the same time. Of course, this event is highly unlikely.

Check with the supplier for the response-time analysis of the EAC system you are evaluating. You are looking for the time it takes to receive an alarm in a busy system. Pay close attention to the definition of "busy" and make sure that if you are comparing systems, the manufacturers are using the same assumptions about loading and what is being measured.

Bottlenecks

Every system has a communications bottleneck. The bottleneck may be large enough that it does not impact the speed required by your system, but it is always a good idea to know where the bottlenecks are in your system so that as the system grows you know the areas that need to be addressed first.

To locate the bottleneck in your system, look for the device that has a combination of the largest number of connected devices and slowest data transmission rate. Sometimes locating the bottleneck is obvious and sometimes you will need a communications professional to help you.

The communications portion of the software is also responsible for monitoring the communications links and the data being transferred to insure that the communication link is secure and the data is correctly delivered.

An alarm should be reported if the supervisory computer unexpectedly loses communication with any of the control panels or other devices it is monitoring. The system should then continue to monitor that node to detect when communications is re-established.

The communications protocol determines the proper delivery of messages. This includes verification that the message was received, and that the message had not been corrupted by electrical noise or other problems.

If the message is not correctly received the transmitting node will re-attempt to send the message. Re-transmissions can be typical because of the many ways in which communications can be interrupted or corrupted.

The communications software will attempt multiple re-transmissions before reporting the error to the operator. Sometimes the level of sensitivity for re-transmission rates can be adjusted. This is helpful when troubleshooting a system, or when you simply want to reduce the number of alarms in a system that is known to have poor communications connections.

Redundancy and Security

Early in this chapter we discussed graceful failure. In a security system it is better if the system does not fail at all. An electronic security system is only as good as it parts. When a single part fails, it can cause failure throughout the system.

The point of redundancy is to make sure that this does not happen. The more redundancy, or duplication, you introduce to your system, the more fail-safe it will be.

Redundancy Tips:

- Protect the power supply of the supervisory computers with a high-quality, uninterruptable power supply (UPS). Make sure that this UPS also protects the computer from the power line surges, spikes, and dropouts that can occur due to lightning or heavy industrial equipment. Some areas are more susceptible to these problems than others. Consult your electrical power vendor to get recommendations for your location.

- Most control panels have battery backup, but you will want to make sure that battery will provide the type of operation you are expecting for the time period you need.

- Most systems consume power much faster if they must support card readers and open doors than if they are only expected to maintain a record of alarms that occur.

  Determine the amount of time you are likely to be without power, the operation you want to support while power is out, and then consult with the supplier of your system to determine if the battery backup is adequate for your needs.

- Make sure your computers and control panels operate independently of one another. Turning off one device should not cause the other devices to stop operating.

- Have a replacement available for the supervisory computer. Computers can fail for any number of reasons, mechanical or electrical. Make sure the replacement computer is an exact duplicate of the original with the same software and add-in cards. Don't forget to upgrade the replacement when the originals are upgraded with either new software or hardware. Check the replacement periodically to make sure it works.

- Automatic backup transfer is available with some computer systems. With this type of system there are actually two computers performing the role of the supervisory computer. Only one of the computers actually issues commands, but in all other respects they perform the same operations with duplicate data bases. The backup computer monitors the operation of the primary computer and if it detects a failure, it assumes command of the system. Making this backup function completely automatic is complex and expensive, but there are a wide variety of partial implementations of this idea.

- Make regular backups of the data in your supervisory computer. You don't backup your data? Well, some people just have to learn the hard way.

The rule is that you must backup as often as you can afford to lose data. If you can't afford to lose more than a day's worth of transactions then you must perform a backup every day.

Don't make the mistake of always backing up your data on the same tape or disk. Rotate the backup media. If you are performing a backup every day, we recommend that you have 7 to 14 sets of media. Perform the first backup on the first set. The second day use the second set and so forth until the 15th day when you again use the first set.

With media rotation you suffer minimum loss should a backup not work for some reason. On a yearly basis make sure a new set of media is rotated in. Most backup media have some mechanical components that do wear out over time.

- On a monthly basis one set of backup media should be archived permanently. This is to protect against the infection of the system with a virus that is not detected until the entire backup rotation has been completed. With an inventory of backups in storage, you can always go back in time to a non-affected copy to re-install to your system.

- Periodically check your backup by restoring the data from the backup. Nothing is more disappointing than faithfully following a backup rotation only to find out, when you need it most, that the data was not being recorded correctly!

- Keep a good stock of new cabling ready for use. Never save old or suspect cabling.

- Use multiple ways of contacting authorities: cellular and land-line phones, modem, or direct link.

- Keep a list of people who can repair your system including first, second, and third choices. Clearly indicate any area of specialization assuming that you may be on vacation when the need to call someone arises.

- Place tape recorders in strategic places. This provides a way for your guards to quickly record their impressions of emergency situations, rather than relying on their memory and writing ability to provide an accurate record after the event calms down.

And, most important, take good care of your staff! They are the best insurance you have for a smooth running operation.

System Design Options

Chapter 9 *System Integration* discusses the design of the EAC system from the security perspective. The prime directive of design is "form follows function." Therefore, the system should be first defined in terms of the function; what is the goal of this system? When the directions in Chapter 9 have been followed and the goal of the system is clear, then the technical detail must be defined.

It will be helpful if you know the following when sitting down with an EAC supplier to discuss the design of your system:

- The number and types of credential readers.

- The number of people that will need access to each entrance.

- The number and types of alarms to be monitored.

- The location of the entrances to be monitored and the distances to the wiring closets or locations where control panels can be mounted.

- The distances between the locations to be controlled and the supervisory computer.

- The manner in which credentials are to be created.

- This is especially important if badges containing photos or signatures are to be created.

    In this case there are some further questions:

    > Will the people needing badges go to a badging station or will the badging station go to them?

    > If multiple stations are being used to create badges, can the resulting data bases be merged?

    Even if personal badges are not created, the designer must decide how the credentials are to be assigned to the people needing them.

- The number and location of operator stations. Some follow-up questions might include:

   Where will the supervisory computers be located?

   Will they need CCTV control?

   What types of connections are available for the computers to share data?

- Take particular note of any routers, bridges, or gateways between supervisory computers on a network as these sometimes affect the protocol that can be used.

- Availability of common carriers. When you have to use external lines to communicate with remote locations, you must determine what services are available.

If you do not know all these items, your supplier will help you determine the answers, but these questions must be answered before an effective system can be designed and installed.

Selecting System Components

A few final notes about some system components that were not covered elsewhere in this chapter:

Printers

When choosing a printer for your system be aware of the fact that while laser printers provide fast, clear printouts, they do not print until they have the image of the entire page in memory.

If you use a laser printer to log alarms they will not print until an entire page of alarms is accumulated or the software package sends a page feed instruction to the printer. Ink-jet printers and dot-matrix printers print a line at a time. Since laser printers must accumulate the whole page for printing it is likely that you will need to upgrade the memory in the laser printer in order to print complicated graphics like maps or photo images.

Monitors

Monitors should be large with high resolution and a fast refresh rate (65 Hz or higher). The monitor is your window on the system status, a small monitor with low resolution will give you a tunnel vision. Today's GUI interfaces, graphical maps, photo identifications, and on-screen video demand high quality monitors.

Backup Media

Tapes are preferred over floppy disks and DAT (digital) tapes are highly recommended. Tapes are preferred because typically the contents of an entire hard disk can be stored on the tape, eliminating the need for an operator to sit and perform the annoying task of pushing floppy disks into the drive one after the other.

DAT tapes are preferred because of their reliability. Removable hard disks or copying the files to another hard disk on the network are also fast reliable means of performing a backup, but these methods typically make it difficult to rotate the backup set and the security of multiple backup sets is more important than convenience.

Computer

The model of computer and type of operating system you should get can become as emotional a subject as politics or religion. Avoid the hype and depend on the good reputation of your supplier. Depend on the security expert that understands your needs and provides good service rather than the latest magazine article. After all, will the author of the article come and fix your system when you have a problem?

You can probably find a lower cost computer in a mail-order ad, but the suppliers of EAC systems typically offer the computer bundled with the system because it is not possible to test every brand and model of computer that is being sold.

New models become available every few months and even the existing models change at the whim of the suppliers. The differences between models can be subtle and in some cases, cause problems with the operation of the system. If you are a computer expert and want to take a chance, you may save a few dollars, but we recommend that you select a supplier and hold that supplier responsible for the entire system, computer included.

Credentials

The type of credentials you choose will determine the type of credential readers and possibly the type of control panel, although many control panels handle a wide variety of credential readers. The choice should be made based on the following:

- the convenience to the users,

    you have to work with these people.

- the replacement cost,

    credentials get lost, stolen, and damaged.

- the reliability of the credentials,

    do you need temporary or long term credentials?

- and the amount of protection needed.

    As indicated in Chapter 3 Credentials, some credentials are more secure than others.

Control Panels

Besides all the features and whatever else has sold you on a specific system, find out what it is going to cost to support one more door. It is amazing how the small system that was so difficult to justify, absolutely *must* be expanded once it is installed and operating properly. Won't you be embarrassed if it costs as much to add an additional door as it cost to put in the whole system?

Cabling

Cabling comes in all grades and costs. It is very tempting to choose the lowest cost alternative because, after all, it's just wire. All cables are not created equal. Pay attention to the cable specified by the manufacturer. If there is some question about substituting a lower cost alternative, ask the manufacturer to evaluate the choice and tell you the effect of its use. Pulling wire is expensive, but finding a problem with bad cables and then pulling wire the second time will cost more than anything you save on the cheaper cable.

Chapter Review - System Design

A system is . . .

> a collection of components that function together for a single purpose.

All EAC systems should have at least two things in common:
1. Sized to meet the needs of the intended function.

2. Safeguarded against failure.

A typical EAC system is comprised of:
- Control panels
- Output devices
- Input devices
- Supervisory computer
- Software
- Communications devices

The terms *client* and *server* refer to computers. They mean that . . .

> the client computer requests service (or data) and the server computer supplies that service.

A control panel in an EAC system is . . .

> a special purpose computer.

In a network, proprietary computers, such as control panels, require that . . .

> all panels be built by the same manufacturer.

A *node* is a generic term that refers to . . .

> any device connected to a network.

A multi-drop communications system has the following characteristics:
- It is easy to add or remove a node from the network.
- Repeaters may be needed to rebroadcast and strengthen the signal.
- When a node fails, only that node is lost.
- When the cable is cut, all nodes downstream from the cut are lost.

A daisy-chain communications system has the following characteristics:

- No repeater is needed because signals are continually rebroadcast.
- The loss of a node is the same as cutting the cable.
- If the cable is cut, the least effect would be the loss of communications to every node downstream and the loss of communications from every node upstream.

Serial ports on personal computers, workstations and most mini-computers use the . . .

RS-232 standard.

Serial ports can be added to a personal computer through a . . .

port expander device, which consists of a card that plugs into the PC's main circuit board.

Modems are used to . . .

create temporary connections to control panels at distant sites.

When setting up an EAC network, it is highly recommended that it be . . .

separate from other computer networks.

The EAC supervising computer should always be . . .

different from the supervisory computer in the corporate or enterprise-wide network.

Specifications for cable and device selection are important because . . .

if you don't follow them, you will have a hard time troubleshooting your system.

Software for EAC systems has three components:

- user interface
- data base
- communications capability

Chapter 8

The following terms are commonly used in EAC systems:

- *Shunt an alarm:* Ignores the alarm input.
- *Energized* and *de-energized:* Changes the state of an output. What happens after the state is changed depends on your setup.
- *Pulsed:* Temporarily changes the state of an output, then returns it to normal.

EAC determines when and where a person can access the system through assignment of an *access level.* The access level contains:

- *Time zone:* defines the times and days access is valid.

- *Access area(s):* defines specific reader locations where access is valid.

Time zones, which can be quite complex, are given a label so that . . .

they can be easily identified.

A data base is . . .

a table of information that can be accessed and sorted in multiple ways.

A problem with maintaining a data base is . . .

that it can get damaged upon loss of power to the computer.

To safeguard your data base . . .

make backups, use an *uninterruptable power supply* on your supervisory computer, and know how to rebuild a data base in case of emergency.

Determining the overall speed of data transmission is a combination of:

- Transmission speed of the system's components.

- Length of the messages.

- Routing ability of the devices that control the communications.

A *real-time event* is one that . . .

is seen on the monitor immediately after an alarm has occurred, not some minutes or hours later.

Real-time, in an EAC system, is determined in terms of human response time in . . .

tenths of a second.

To locate a bottleneck . . .

look for the device that has a combination of the largest number of connected devices and the slowest data transmission rate.

Redundancy throughout the system is important because . . .

the more duplication you introduce to your system, the more fail-safe it will be.

System component selection considerations are:

● *Printers:* Be aware that while laser printers provide fast, clear print-outs, they do not print until they have the image of the entire page in memory. Line printers print continuously as long as they are being fed data.

● *Monitors:* Should be large enough, have high resolution and a fast re-fresh rate of 65 Hz or higher.

● *Backup media:* Tapes are preferred over floppy disks and DAT (digital) tapes are highly recommended.

● *Computer:* Follow your supplier's recommendations.

● *Credentials:* Should be convenient for users, have a reasonable replacement cost, be very reliably read in readers, and provide the amount of protection needed.

● *Control Panels:* Besides all the features that sold you on a specific system, find out what it is going to cost to support one more door.

● *Cabling:* Pay attention to the cables specified by the manufacturer. If there is some question about substituting a lower cost alternative, ask the manufacturer to evaluate the choice and tell you the effect of its use.

Chapter 8

QUESTIONS

1. What are the three components of EAC system software?

2. What is the difference between a client computer and a server?

3. What are two methods for connecting nodes to a Control Network and what is the result of a failed node in each?

4. Name at least three methods of adding redundancy to a system.

5. What is an access level?

Chapter 9

System Integration

A Security Perspective

By Joseph Barry, CFE, CPP and Patrick Finnegan, qp

Benefits of an Integrated Security System

The ultimate purpose of any security system is to counter threats against assets and strengthen associated vulnerabilities. To put it another way, "the weak and vulnerable shall be made strong." Part of that strength comes from record keeping that illuminates past events as well as from devices that survey and/or control current conditions.

In the not too distant past, historical records were generated through unrelated systems, including handwritten documents. This resulted in training people to adapt to a variety of conditions, including unrelated software, and finding records about a single event in unrelated databases and file cabinets. The inefficiencies of this type of system caused information to slip through the cracks.

An integrated system controls and/or monitors current conditions and keeps historical databases in interrelated computer files. This provides a cost-effective way to administer complex security functions.

The focus of this chapter is to help you plan an integrated system from a security perspective. Without good planning, the success of the system is a matter of trial and error. During the trial, of course, you risk the error of weakening security . . . something you would definitely want to avoid.

Creating an integrated system is not an exact science because it uses devices manufactured by competing companies. In many cases, these devices use state-of-the-art technology that maybe unfamiliar to the user at the time of purchase.

Through the use of a solid plan, the system designer can stay on top of the facility's needs, educate him or herself on the latest technology, pinpoint objectives, stay within budget and design an excellent integrated security system comprised of a wide variety of parts.

The difference between an integrated system and interconnected devices is a matter of control. That control is provided through *integrated software* that resides on a single, supervisory computer (although it might be administered throughout a network of computers and/or computer and transponders).

> *Example:*
> Although a security system may contain a CCTV and alarm system that work together, if those two systems are hooked to separate controllers (or no controllers at all), the relationship between these devices is **interconnected.**
>
> If, however, the CCTV and alarm system share the same controller and the software that controls one device is the same that's used to control a completely different device, that system is **integrated.** All the devices in the system generate historical records that reside in related files.

Well designed integrated systems save:

- **Management Time**

  Record keeping is easy. No fumbling for information is required.

- **Employee Training Time**

  Employees only have to learn one set of software and operational response rules.

- **Response Time**

  Procedures are well thought-out because they are related.

- **Physical Space**

  Reduces the need for multiple monitors.

- **Money**

  Fewer people can do more work.

Determining the Scope of the Plan

You need an overall plan that defines what you want to accomplish. The lack of a comprehensive plan usually results in cost overruns and failure to meet your primary mission, which is, of course, to counter threat and keep the facility safe.

No matter how good the parts might be, without a plan, the sum of them will not work. An excellent design for a complex integrated access control and alarm system can be rendered useless if it fails to account for the architectural design of the building in which it is housed.

Throughout the planning process, your primary focus should always be on the protection of assets. Are those assets a:

- Single item?

- Group of items?

Do those assets reside in a:

- Single facility (building or portion of a building)?

- Group facility (a number of buildings or several different portions of one or more buildings)?

- Site (a number of buildings and support structures)?

As you progress, you'll look at your assets from many different points of view. The more area you have to protect, combined with the degree of threat to the assets contained in that area and the vulnerability of the area's perimeters, the more complex the scope of the plan.

Examining Current Systems: The ideal situation is to plan a security system for a new construction. This lets you select all new components, which minimizes potential conflicts between hardware and software.

If, however, you have to enhance or upgrade a current system, you must decide whether you are going to replace all of the existing security components or attempt to salvage some of them for your new system.

> **Of Minimum Concern:** Usually, currently installed sensors and electronic door strikes can be integrated with new products.

Of Maximum Concern: If you decide to use parts of your currently installed alarm system, CCTV system and access control system which is not now integrated into a single operating unit, you face complex concerns. You must consider:

Were the existing components designed to be operated through a computer controlled system?

With older systems, you usually will not achieve full integrated control without major modifications. Many times, these modifications are not cost-effective.

Older alarm systems, for example, were not designed to work within computer-controlled situations. Their dry contacts are able to respond to remote alarms, but they are not designed to pinpoint the zone or sensor location that is doing the reporting.

Deciding whether to retrofit old technology depends on what you expect from the final system. Keep in mind, however, that as technology ages, there are fewer and fewer repair people available who understand it, let alone have the ability to fix it.

Is your existing system set up on a proprietary basis?

A proprietary system uses a single source of supplies. Generally, that source is a single manufacturer and its product line.

If you plan to use a competitive bidding process, sticking with a proprietary system greatly narrows your options.

It is, however, possible to construct an integrated system using equipment from different manufacturers. The problem with this approach is that when something goes wrong, no one will take the blame. This is especially critical when new technology is matched with old.

Do you need special agency approvals?

Not all integrated systems have Underwriters Laboratories approvals, even if portions of them might. This can create problems with fire departments and insurance companies, depending on their individual requirements.

Check with regulatory agencies before you start to upgrade your system and incorporate their requirements into your plan.

The result of this initial thought-gathering process should help you decide whether you need to start from scratch or keep some existing components. If you have a proprietary system and want to stick with it, you must decide whether the manufacturer's updated product line is broad enough to meet your current and future needs.

Involving Your Managers and Staff: No matter how much you or hired experts might know about security, it is essential that top management and key staff members be kept informed of the planning processes.

Management is responsible for budgets and consequently, must be kept clearly aware of the advantages and disadvantages of your project as it progresses. By keeping them informed as you go, you avoid hitting them with too much information when it comes time to make final decisions.

Justifying expenditures is a very important issue in an era of downsizing and budget constraints. You must present your budget justifications in a way that stresses return on investment. Too often security specialists see their activities as charges against profit, rather than helping the organization achieve its overall financial goals.

Positive presentation of the overall benefits of the plan to your company's profit margin will not only assist in getting the project funded, but will help you defend it, should cuts be suggested.

The supportive buy-in of other department managers is also essential. Keep in mind that there are a number of ways to go about a project, with no one being more right than another. Department managers know their job and the facility in ways that may differ from the way you view these things. They can best describe assets and the costs of replacing those assets should harm come to them. The more they offer, the better prepared you are.

By having the support of department managers and staff, you minimize complaints at the end of the project, or the charge that you overlooked assets that should have been obvious.

Reviewing Products and Options

Although making product selections before you perform an analysis is foolhardy, knowing what is available, including approximate costs, will probably open up your line of inquiry.

Now is the time to bring in a system design consultant. This person will steer you to products and help you formulate the right questions. Select the design consultant by checking his or her existing customer base and getting recommendations. Look for an affiliation with the American Society of Industrial Security (ASIS) and/or the American Institute of Architechs (AIA).

Based on your new-found insight, attend trade shows, request literature, read articles and talk to salespeople. Education takes time and consequently, you don't want to be overwhelmed with options that all have to be weighed toward the end of the process.

Technology is changing the way things are done by the minute. Today, it is quite common to discover systems that do things that surpass your wildest dreams. Even though you don't know exactly what you need, you probably can generalize enough to get good information.

> *Example:*
> Are you considering integrating a motion-activated lighting system with retractable gates, card readers, CCTV cameras and sensors in a parking garage? What products might complement this effort? Who are currently using these products? Can you get a list of people who set up projects similar to yours? How do these people anticipate the future and potential changes in technology?

This is also a good time to check on the reputations of local installers and the types of equipment they support. Falling in love with specific technology at a trade show is no good if there are no local installers capable of maintaining that technology. On the other hand, a local installer with an excellent reputation might service a limited line of products. What are those products and will they fit your needs?

If your new system requires communications, what lines will you use? Standard telephone links? Custom digital carriers? Cable networks? Wireless networks? A combination of network types? Get an overall view on communications so that you won't be overwhelmed with terminology that is more common to computer technologists than it is to security personnel.

If you will be sharing communications (such as a cellular service or telephone lines), find out if there is the potential of line jamming during peak communications hours. What might happen to your data transfer (including remotely reporting alarms) if transfer rates slow down or stop altogether?

Let your past experience be your guide, but keep an open mind. It is up to you to link the latest in integrated technology with your goal of protecting assets. Expect surprises as a matter of course.

Assess the Threat by Making Lists

Before you can design a security system, you need a clear picture of what it is that you are protecting and what constitutes a threat to those things. This is called a "threat assessment."

While there is no set way to determine threat, careful planning, articulation of concerns, involvement of managers and staff, and reflection goes a long way toward being perfect.

Every facility is unique, not only because buildings are different, but because the assets they contain and the value placed on those assets differ so greatly.

A security system that works well for a bank probably would not work with the same effectiveness for a shopping mall. Systems that protect nuclear weapons would be too expensive for a supply room in a fast food restaurant.

In order to examine a system, you must put a value on the items to be protected. It is this value that determines the cost of replacement in the event of theft or other loss.

Using the base value as a guide, a security system that at first seems expensive might actually be quite reasonable when compared to replacement costs.

The following lists will get you started:

> **Asset:** List your assets. Make the list as comprehensive as possible.

> **Asset Value:** Place values on each asset. These values represent the original cost and replacement cost. Replacement cost includes the actual replacement expenditure, plus lost employee time, lost profits and other costs affected by the loss.

> **Asset Location:** Cite the location of each asset. This is important because the same asset or asset group might require more or less protection depending on where it is kept. Money changing hands on a bank counter, for example, requires a different type of protection than money that's stored in a vault.

> The location of an asset is tied to the skill level of your adversary. If the item being protected is located in a vault, which itself is inside a compound that has excellent physical barriers (fences, walls, alarms,

etc.), then the amount of skill and **time** necessary to obtain the asset increases.

> **Tip:** The longer it takes to get to an asset, the more likely it is that a thief will be deterred by fear of detection.

Mission of Location: Cite what each location is used for. A public location, such as the shopping area in a store probably contains the same type of assets as does the store room. The mission of these two areas, however, is entirely different and the areas are vulnerable in different ways.

Potential Adversary: List and describe who would want the item being protected and how skillful he or she might be at getting it.

> *Example:* A thief with no knowledge of security systems or safes would have a very difficult time stealing from a bank vault, but the same thief might have no problem breaking into a warehouse.

While there are many more lists that can go into threat analysis, they will suggest themselves as you refine your lists, starting with the ones recommended here. Very often the descriptions on one list lead to questions that are answered by developing another list.

During this examination process, you will begin to develop your final objectives.

Start by examining small areas, then work into bigger ones. Leave grouping for last. You don't want to ignore any area, no matter how remote.

If your facility is spread out over a number of buildings, you should complete an assessment for each building, then draw all the assessments together for final consideration.

This process will cause you to group and regroup your information in various ways, leading to numerous subcomponents that might not have been initially obvious.

All of this information will provide the backbone for the next steps.

Formalizing Your Analysis

You must study your analysis in detail in order to thoroughly understand its components. As you do this, ideas about your needs will emerge that provide insight into improved efficiency, protection and cost-effectiveness that can only be met through an integrated system.

At first, the size and scope of your project might seem unmanageable. By addressing smaller parts, you should be able to:

1. Manage information flow.

2. Identify issues that might become problems.

3. Control the direction and timing of the development stages.

4. Monitor and control the project's development.

5. Handle large, complex issues and tasks in a controlled manner.

You formalize your analysis through a **Security Concept Plan**, which consists of a:

- Requirements Analysis and System Definition Plan

- System Engineering and Design Plan

This master plan defines your entire project (i.e., protection goals, strategies, systems, threats and countermeasures). It is your project's map and eventually will be used to measure your project's success.

Typically, this plan is fluid and is continually adjusted as you become more and more informed. Creativity and an open mind are required during this stage. The effort you put into this plan will pay off in its success later on.

Matching the dream of system integration with the reality of selecting products from unrelated vendors requires concept adjustments. The clearer you express your needs, the more you will be able to make the right product choices.

- Security Concept Plan -
Requirements Analysis and System Definition Plan Sections

The *asset* is always the focal point of each analysis. Multiple assets must be evaluated individually in terms of threats and protective measures and therefore, must be structured individually.

The *mission* of the area in which your asset is kept might change from site to site within the boundaries of the facility or facility's holdings; consequently, methods of protection can differ even though the asset remains the same.

Defining the areas to be protected might seem difficult because of all the required detail. By being patient and methodical, this difficulty can be overcome. Define everything. Save grouping and summarizing for very last.

Draw on your well-developed pre-planning lists to help you complete the following sections in the **Security Concept Plan**:

> **Asset Definition Section:** Define the assets that require protection, including original and replacement costs, location, etc. Determine how critical that asset is to the mission of the area in which the asset is kept.

> **Threat Assessment Section:** On an asset-by-asset basis, describe threats and potential adversaries that could cause a loss.

> **Vulnerability Assessment Section:** On an asset-by-asset basis, determine your ability to adequately respond and/or counter a potential threat. Include your current capabilities, controls and mission requirements.

> **Site Survey Section:** On an overall basis, examine your territory. Cite specific problems and requirements. Relate this survey to your assets.

> **System Requirements Analysis Section:** Based on your analysis, define your objectives. What level of protection do your assets require and where and how is that protection going to be deployed?

- Security Concept Plan -
System Engineering and Design Plan Sections

Your goal during this phase of planning is to create a plan that provides protection in depth.

Compromises are not made at this time. At this point, you must define and specify the very best security system possible.

The following requirements will help shape your plan. You probably will have additional requirements, based on your previous analysis. Always remember, your situation is your guide. Everything else is advisory.

> **Hardware and Software Requirements:** Evaluate and select equipment that maximizes reliability and detection probability. Determine whether the equipment you are considering can be integrated or not.
>
> If you are required to retrofit an older system, the issue of compatibility is also essential. Keep an open mind about cost and effectiveness.
>
> **Personnel Requirements:** Keep your system user-friendly. Determine how many people will be required to monitor, manage, maintain and respond to this system.
>
> Make sure that your security force does not lose confidence in the system. If it is not highly reliable, they will ignore it. If it is too complex, they will avoid it. (Remember, if there is no response to an alarm, for whatever reason, the system is ineffective.)
>
> **Operation and Technical Procedural Requirements:** Document what procedures will be required to maintain the system. This includes sections on operators, maintenance and supervision, among other things.
>
> **Support Requirements:** Clearly define what you need to support the system. This includes setting up specifications for leased communication lines, power outlets, employee hiring and training, etc.

Evaluating and Selecting Components

Once you set forth the very best plan possible, and document the reasons for your selections, it is time for you to determine whether it truly meets your needs.

At this point, you need to adjust your goals, budgets and selection decisions based on a wealth of the information you've collected. This evaluation process includes:

Threat and Vulnerability Assessment: Does the proposed system counter threats and vulnerabilities that have been previously defined?

Concept of Operations: Does the proposed system require operations that are realistic. Does it satisfy the basic needs of the facility?

Economic and Other Constraints (yes, dollars matter): Is the proposed system economically sound and offer a good return on investment? Is it cost effective? Does it allow for expansion in the future? Can it be upgraded to take advantage of new technologies?

Operations Requirements: Can it do the job in an effective manner? Does it conform to and support other facility operations. (Warning: If security impedes other facility operations, then eventually security will be scaled back or suffer other problems. The plan must mesh smoothly with general facility operations.)

System Requirements: Does it conform with the System Requirements Analysis? Will it provide protection for identified assets? Will it counter identified vulnerabilities?

The evaluation of these factors enable you to determine whether your design is adequate to meet perceivable threats. This evaluation will also provide the justification for expenditures, which you'll set forth in your formal specifications.

Setting Specifications and Getting Bids

Once you've settled your requirements, you are ready to draft your final specifications and start taking bids. At minimum, those specifications should include:

- Equipment requirements

- Installation services

- Fee-based training services

- Ongoing maintenance services

- Communication services (optional)

Do not be surprised if specified system changes during the bidding process.

Keep an open mind. Often dealers and installers are aware of aspects of projects that might not be readily apparent to individual buyers. On the other hand, your prior research should guard you against being pushed into accepting unneeded equipment. Working with a system design consultant will further buffer you against a last-minute oversell.

Always include the cost of training in the installation contract. The dealer (possibly together with the equipment manufacturer) must provide training on a fee basis. Before you finalize the contract, check to see that the training programs offered are effective and respond to individual needs. Talk to former students in order to make your evaluation.

> *Remember:* Installation and maintenance manuals do not provide enough information on how to properly operate equipment. The best way to learn complex tasks is to have other people show you how things are done.

Implementing the New System

Once you accept bids and close contracts, your next round of work begins. In addition to preparing work schedules, you must determine how normal operations will be affected during installation and make appropriate adjustments.

Implementation includes:

Prepare Facilities: Set delivery dates for the equipment, installation and new furniture (if required).

Install the System: Although installation is usually done with trained technicians, security personnel should monitor the installation process. Decisionmakers should be available to make minor changes, such as relocation of sensors, etc. Expect some problems and be ready with solutions.

Test the System: This is extremely important. Every sensor, camera, access point and communications link must be checked. Consider documenting this process through the use of photos and/or video recording.

Train the Operators: Keep your operators informed throughout the process. Ensure that they understand how to use it once it is running and that they can respond appropriately to a variety of situations. Make thorough use your dealer's training programs.

Train the Users (if needed): A new system might require more involvement and/or awareness from the facility's staff than the old system. Make sure that everyone understands what is in place, where they can get help and how to avoid problems.

Establish a Support System: Create maintenance schedules. Store spare parts. Identify an emergency service.

Reviewing Ongoing Operations

The system and your facility are a living entity. They will grow and change together and, quite frankly, the only thing that is constant in today's world is change. Keeping up with it is a requirement!

Continually use the analysis instruments described earlier in this chapter to keep your vulnerability assessment in focus. Use these assessments to ensure that your system remains current.

Typically, an integrated system provides security management with the ability to pool resources, leading to a more flexible approach to staff deployment. This might mean fewer people will do more things, or newly hired people will have greater responsibility.

The final measure of the success of your project is whether it meets all the requirements that were laid out in the plan. In the beginning, your objective was to protect assets. You will have achieved a truly integrated system when you can prove that it:

Protects all critical assets

Incorporates a valid and complete threat assessment

Reduces vulnerability

Is compatible with facility operations

Provides cost/performance effectiveness

Chapter Review - System Integration

The difference between an integrated system and interconnected devices is a . . .

> matter of control. An integrated system is controlled by a single, supervisory computer.

Well designed integrated systems save:

- Management time

- Employee training time

- Response time

- Physical space

- Money

Before you start planning your system, you must . . .

> determine the scope of your plan.

Determining the scope of your plan involves evaluating:

- Assets

- Facilities

- Current systems

- Need for special agency approvals

- Staff input

The lists useful in evaluating potential threat include:

- Assets

- Value of assets

- Location of assets

- Mission of location

- Potential adversary

The *Security Concept Plan* consists of . . .

> a "Requirements Analysis and System Definition Plan" and a "System Engineering and Design Plan."

Recommended sections to the "Requirements Analysis and System Definition" portion of your plan are:

- Asset definition section

- Threat assessment section

- Vulnerability assessment section

- Site survey section

- System requirements analysis section

Recommended sections to the "System Engineering and Design Plan" portion of your plan are:

- Hardware and software requirements

- Personnel requirements

- Operation and technical procedural requirements

- Support requirements

Evaluating and selecting components consist of:

- Threat and vulnerability assessment

- Concept of operations

- Economic and other constraints

- Operations requirements

- System requirements

Implementing the new system includes:

- Preparation of facility

- Installation of equipment

- Test of equipment and all aspects of the system

- Train operators

- Train other users (if needed)

- Develop support systems

QUESTIONS

1. Why is an integrated system valuable?

2. Name some issues involved in determining the scope of your security plan.

3. Name as many lists as possible to help assess the threat to your assets.

4. A formal analysis of your security needs is called a "Security _____ Plan."

5. Name at least four activities required when implementing a new system.

Chapter 10

Case Histories
Noteworthy EAC Installations

Now that you have a good idea about how electronic access control (EAC) systems work, this chapter provides stories about noteworthy applications.

The stories we've chosen emphasize that no two security systems are designed for the same reasons. With that in mind, most of the people we interviewed would be pleased to answer serious inquiries about the issues they raise.

We do not mention product names, however, even though we learned about many excellent components. We hope that these stories promote interest in discovering what all EAC manufacturers and dealers have to offer, rather than limiting you to any one source.

Of course, EAC is only as good as the staff that makes it work properly. Everyone interviewed for this chapter talked about the people who run their systems. Unlike the traditional night watchman who may not have seen his duties as a profession, many of today's security officers must:

- Greet visitors, making them feel comfortable, even while screening their backgrounds and directing those who accidentally wander near controlled areas.

- Maintain computer records and analyze history logs.

- Understand and maintain complex electronic equipment, such as CCTV, credential readers, intrusion detection devices and electronic locks.

- Understand electronic communications, wired and wireless, and know how to use alternative sources should lines fail.

- Understand and enforce numerous Federal and local laws and regulations.

- And, possibly most important of all, risk their lives when emergencies occur.

Jackson Memorial Medical Center

University of Miami
Jackson Memorial Medical Center
Miami, FL

Downsizing is not an issue for the University of Miami Medical Center (UMMC). During the decade that Tony Artrip has been Director of Security, UMMC increased his responsibilities by almost 300%.

UMMC is ranked by physicians as one of the nation's top 25 medical centers. With approximately 1,600 licensed beds, it is one of the nation's largest, busiest and most technologically advanced healthcare centers. A map is available from UMMC's public relations department to give you an idea of the scope of its facilities.

When Tony began in 1985, he oversaw 16 security officers with no electronic monitoring support. Today, he manages 60 security officers, at least 260 EAC card readers, over 800 individual alarm points, and 100 CCTV cameras, complete with video recording capability.

During this time, the center, which is a collection of individual facilities, built four new buildings, purchased two more and it is continually negotiating office space from other buildings in the surrounding area.

Growth demanded that the security department resolve three problems:

- Reduce theft in offices and patient's rooms.

- Increase the feeling of comfort and confidence by people using the facilities.

- Eliminate trespassing. This trespassing usually does not have criminal intent. As UMMC is open to the public, visitors often wander into unauthorized areas by mistake.

Tony credits his department's development and success to good planning and strong support from UMMC's upper management.

Upper management insists that prior to building or leasing a new facility, all involved department heads must meet with the Security Department. During this time, department assets are identified, floor, ceiling and wall plans are analyzed, and EAC components are selected. Nothing proceeds until the security planning is complete.

Each department is granted three time zones. The first is for 24 hours, the second usually represents a shift that runs 7 days a week and the third is customized around normal business hours and holidays.

Detailed records of everyone using the access system are kept. Individuals are assigned their own *custom access level*, identifying the times and days when he or she may use specific doors. By maintaining custom levels, access is controlled by *real need*, rather than general category. No one is given more access than is required.

Proximity cards are used to keep staff traffic moving swiftly through access points. These are clipped to photo ID badges, which must be worn at all times.

Incredibly, 95% of the 60 uniformed security officers are qualified to work the EAC system, something that is not true of other large organizations.

To learn the system, new officers are paired with senior officers during second shift. Passing probation requires that every guard memorizes the location of all alarm points and access areas throughout the facility. All guards are trained at an EAC monitor and most are able to issue commands during an emergency.

Tony particularly likes the fact that alarm points and the access system are integrated. Although the CCTV system is activated through a separate controller, the camera responds to alarm points connected to EAC. When a point is triggered, such as during an access transaction, CCTV cameras swing into service.

How does Tony and his staff keep up with the growth and changes to the system? Tony encourages continuing education and department training. He makes sure that everyone welcomes change and the organization required to make that change go smoothly.

Halifax Medical Center

Halifax Medical Center
Daytona Beach, FL

Within the mid-1990s, the purpose of the Halifax Medical Center (HMC) changed radically. What began as a general hospital emerged into a group of nonprofit health care businesses. These include an at-home nursing service, insurance services (HMO and PPO), a hospice center and, of course, basic medical treatment centers.

Much of this change was brought about by mergers. The result? The Security Department, under the direction of Mitch Norton, is now responsible for over 50 locations. Of that, all the largest facilities have EAC, with several hundred readers in place.

The biggest problem in developing system-wide EAC for newly combined businesses was getting people from diverse backgrounds to agree upon access requirements. Before the mergers, people were not greatly restricted. After, inter-facility access rules had to be imposed.

According to Ethan Lewis, ADT Security Systems, another problem was blending a diverse group of electronically controlled applications into an integrated electronic security system.

Before choosing their current system, the medical center hoped that an installed heating, ventilating and air condition control (HVAC) system could be updated to manage access. After some trials, management discovered that the HVAC system was not specific enough for security. The dream of total integration supervised by a single computer meant too many trade-offs that were not in the best interest of anyone.

Once security needs were firmly identified, the search began for the "perfect" system.

Technology changes. The "dream" EAC system Mitch Norton wanted was not yet ready for mass production. Gambling on the future, he participated in beta testing this new EAC system and was rewarded for his efforts.

Beta testing means that users have a strong say as to what goes into software design. Unlike using off-the-shelf products, which may work well, but may not be technically up-to-date, beta systems are shipped with problems meant to be overcome through on-site testing.

While overcoming problems can be hectic, Mitch strongly recommends participating in the beta testing process. It allows you to:

- Work closely with software and hardware developers and learn what goes on "inside the box."

- Influence the development process, resulting in a "customized" system from what started as a generic product.

- Learn how to test the system. As a beta tester, Mitch learned about communications, line connections, device specifications, product testing and trouble shooting. To his knowledge, there are no universities that teach these things.

The biggest challenge in setting up the system, however, was fighting Florida's high voltage environment. Lightning is so frequent and powerful that surge protection is required on all electronic devices. Mitch's advice to people in similar environments is to treat each device as a special case and individualize surge protection to meet its requirements.

Today, the completed EAC system meets the access needs of a diverse group of facilities and the people within them. By using two types of authentication devices on the same system, everyone is satisfied.

- **Keypad Use:** The medical staff required a system entirely free from ID cards and badges because they often respond to emergencies during irregular hours. *Example: A* doctor might be on a family outing away from home and the ID card when paged to return to the hospital.

- **Wiegand Cards:** These secure and very durable cards are issued to all support staff and are used during normal business hours.

Three conclusions can be drawn from this story:

1. Too much emphasis on system integration leads to overly generalized applications.

2. When you trust the dealer and manufacturer and you want the very latest in technology, take a chance beta testing a system.

3. Give users the type of authentication devices they need. Don't force them to use one kind in the name of convenience.

Lam Research Corporation

Lam Research Corporation
Fremont, CA

Setting EAC system standards and enforcing them around the globe is a big issue for Ed Loyd, Security Director for Lam Research Corporation.

Lam Research is one of the world's largest manufacturer of equipment used to make microprocessor chips. Their customers include Intel, Motorola and Texas Instruments as well as major Japanese and Asian corporations.

In the United States alone, Lam has 18 buildings at its corporate headquarters, spread out over four square miles. They also have major facilities in Massachusetts, Taiwan, Japan and Korea.

To unify and strengthen security procedures, it is necessary to install the same system in each of these localities. Before we tell you about what is involved in achieving global security standards, let's look at some of the unique features of the EAC system installed at Lam Research's Fremont headquarters.

Installation of a new integrated system began in 1994, replacing several unrelated card ID systems. Design objectives included:

- Controlling all credential enrollment from one facility.

- Monitoring all EAC transactions from one facility.

- Creating a time and attendance tracking system for Lam Research contractors.

- Creating tighter control over access areas with antipassback.

Lam Research employs numerous temporary contractors who perform specialized tasks within their manufacturing facility. Before the new EAC system was installed, however, Accounts Payable had a difficult time assessing whether or not these contractors actually worked the hours stated on invoices.

To solve this problem, dealer Jim Mossburg, Information Security Integration Technology (ISIT), had software custom designed to link into the main EAC program. Today, at any given moment, Lam Research can tell when contractors are or are not on the job and automatically associate this information with invoices.

Tighter control over access areas was also achieved through the use of antipassback throughout the system. Today, everyone must check in and out of areas in the right sequence, or an alarm is sounded, their name is displayed on the guard's monitor and their image is captured on video tape. Unfortunately, upper management, who used to freely roam the facilities, had a hard time adjusting to the new system.

According to Ed Loyd, after managers set off repeated alarms, they finally began to comply. Once the managers accepted the system, the rest of the staff followed without complaining.

The biggest problem, which has yet to be solved, is installing the same EAC system in facilities around the globe.

US-based multinational companies prefer to use the same security technology wherever they have facilities. Unfortunately, many EAC systems in Europe and Asia are not available or widely used in the United States, and vice versa.

Multinational US-based companies want to depend on overseas dealers for installation and support. Buying "local" also saves money because stiff import duties and other taxes are avoided.

Unfortunately, regulations around the globe differ from country to country and so do the attitudes of dealers. Consequently, maintaining company-wide standards are very difficult. Lam Research is making strides to accomplish this, however. It currently has a system in Taiwan that communicates over phone lines with headquarters in Fremont, California and is installing another in Japan.

Large Law Firms

Overview Reported by:
Technology Controls
Valencia, CA

Dave Scripture of Technology Controls would like to share with you the names of large legal clients using EAC systems, but he will not because of risks involved. He did, however, give us the benefit of his experiences.

The legal profession, including some insurance companies, specializes in win or lose situations. The problem for security involves how the people who lose cases deal with the outcome. Losing can trigger irrational behavior that fuels the desire to physically harm the winners. When this behavior might erupt, however, is anyone's guess.

To make matters worse, attorneys often advise controversial clients. Firms that provide services to tobacco companies or abortionists, for example, are subjected to protesters, some of whom are violent.

Most people think about security as keeping bad people outside. In this application, however, security is designed to monitor the emotionally unstable people who might already be inside.

Two issues affect this type of security design. The first is that prosperous law office interiors must be kept attractive. The second is that law firms often share buildings with other businesses. This means their own security needs must not interfere with the rest of the tenants, or put those tenants at risk.

Dave reports that the security for one large two-floor legal firm starts on the ground floor of a high rise building. Upon entrance, every member of the firm submits an EAC card for reading. Immediately thereafter, his or her photo ID appears on a guard's monitor and the guard visually confirms identification.

Guards play a key role in this system because their alertness and insight is required to calm emotionally charged situations. Consequently, the bond they establish by identifying authorized users — not only by sight, but by mannerism and voice — is extremely important.

EAC cards also activate use of elevator buttons. Should non-authorized people step off the elevator with those who are authorized, they are asked to leave. If they don't leave, they find themselves alone in a dead-end hall, monitored by CCTV, motion detectors, a metal detector and an EAC checkpoint. The only way out is back on the elevator.

Inside, all doors used in the system are controlled by antipassback with in and out card readers. Without exception, users must "card in" at each access point, then "card out" when leaving. The system tracks every act and an alarm sounds when a mismatch occurs.

Interior doors are also monitored by active infrared (break-beam) sensors, video motion detectors and metal detectors. All video recording is done real-time; no time-lapse recorders are used.

To control crowding, some rooms are limited to a specific number of people. File areas, for example, are restricted to four people. If more than four try to use the room, alarms are sounded. Likewise, if four people enter a room, but only three leave it, after a set period of time, alarms go off.

Panic alarms are placed at reception areas throughout the facility. These are audible and/or silent, depending on department needs. All send signals to the main guard station, where security officers can respond by initiating a complete or partial lockdown within seconds.

Does this type of security setup work?

According to Dave, despite readers, detectors and cameras, the system is unobtrusive. More important, it does an excellent job of immediately identifying the exact location where hostility is occurring.

Such a system installed at an insurance company stopped a threat when a disgruntled customer approached with a gun. Guards, who were familiar with authorized people, spotted him when he was trying to sneak through the first checkpoint. They immediately locked down the facility so that innocent people would not walk into the area, tackled the man to the ground, got his gun and called the police before most people were aware of what was happening.

Montana Power Company

Montana Power Company
Colstrip Project Division
Colstrip, MO

Ken Miller, a ADT dealer in Billings, Montana, had to solve three problems when designing the EAC system for the Colstrip Project Division of Montana Power Company. The first was that all cabling had to pass through buildings producing between 330 and 776 mega-watts of electricity. The second was dealing with minus-30 degree winter temperatures and the third, remoteness.

The Colstrip Project Division produces electricity consumed by Montana and a good portion of the Pacific Northwest. It is located in Colstrip, a town of 3,100 people, which is 120 miles southeast of Billings, a city of over 100,000.

Being located in a very small community means that the plant's security staff has to be extremely resourceful and self-sufficient. There are no repair technicians nearby.

Mike Ames, Director of Security for the Colstrip Project Division, achieved self-sufficiency by requiring that his staff master every aspect of the electronic security system, including wiring, programming and troubleshooting. This increased knowledge became a plus for tight budget considerations.

With increased competition in the electrical utility industry starting in the early 1990s, the Colstrip Project has been to justify its budget in terms of return-on-investment. One way this has been done was to create value-added services. Mike turned his electronic security expertise into a consulting service in which he offers system design, monitoring and troubleshooting to other Montana Power divisions.

Data and video transmission is smooth and fast because the Colstrip Project owns its own microwave network. Transmission speed is similar to that of a T1 communications channel. It allows for clear real-time monitoring from over 200 miles away.

System designers were initially worried about electromagnetic interference, however, it never became a major problem. Prior to installation, every surge protector used in the system was selected from ADT's pretested list of recommendations. This paid off in trouble-free protection and reduced troubleshooting time.

One big problem that had to be overcome was the effect of extremely low outside temperatures. Wiegand cards were the first cards used. Unfortunately, in the winter, when people spent long periods outside, these cards became too cold to read. Consequently, they had to be warmed against human skin in order for them to activate. Replacing Wiegand cards with proximity readers solved this problem. Today, no one has to expose his or her body to a minus-30 degree card in order to gain access.

Of all the buildings, the unmanned substation is of particular interest. It is monitored by photoelectric motion detectors, numerous alarm points (including fire detection), CCTV and a digital video system. All information is transmitted via the microwave network. Despite the remoteness of the unmanned station, the facility is now under close surveillance. In addition to Mike's staff, he maintains an excellent relationship with the Colstrip Sheriff's Department and other law enforcement agencies who all help to check up on things.

Without properly functioning equipment, sharp employees and community involvement, managing the security of this remote large power plant would be difficult. It helps, according to Mike, that, in addition to the technical expertise gained from working with EAC, he comes from a law-enforcement family with ties to sheriff and police departments throughout Montana.

Englewood Hospital & Medical Care Facility

Englewood Hospital & Medical Care Facility
Englewood, NJ

Englewood Hospital is a community complex that delivers specialized care to residents in 40 townships. It has with more than 2,000 employees, 1,000 volunteers, a 450-member medical staff and 700 nurses. Furthermore, it serves 24,000 admitted patients and 100,000 outpatients each year.

Despite its size, Englewood has the distinction of going from average security in 1994 to winning the 1996 Lindberg Bell Award for security excellence in healthcare. How did they make such a rapid transformation?

The medical facility has an "Our Patients First" motto, which they fulfill by implementing the best technology run by top practitioners. In the early 1990s, however, the security department was not keeping up with technological advances. While security officers were well-trained, their equipment was rapidly falling out-of-date and, in many cases, beyond repair.

To improve security, Englewood hired Steve Gaunt, CHPA, CFE, as their security director. He had a history of staying current with technology and, in 1992, he won the Lindberg Bell Award for his previous employer.

Steve believes that excellence in security is attributed to how the healthcare facility regards security issues, a well-trained, friendly staff and the use of the latest technology. He points out that the facility invests in security when it perceives security as an important part of public relations.

No matter what types of surveillance procedures security officers use, these officers are the first people to greet hospital visitors and the last to say goodbye when visitors leave. At Englewood, they also take part in community programs, such as "Kid Safe," a crime awareness program, senior citizen health fairs and Boy Scouts, to name a few.

Good security helps attract new patients. People are vary aware of problems caused by nonrestricted pedestrians roaming a facility. In some cases, the quality of hospital security outweighs the medical staff's ability to use the latest surgical techniques. After all, excellent healthcare requires a healthy physical environment, both of which Englewood has.

Even though Steve's staff now uses the latest in technology (as you'll read below), he believes that hiring officers who are serious about their profession is critical. He strongly recommends that hospital security officers join the

International Association for Healthcare Security & Safety (IAHSS), which is the association that created the Lindberg Bell Award.

Founded in 1968, IAHSS is now the largest association of its kind. It works to improve and professionalize security in medical care facilities through the exchange of information between its members. (See the *Appendix* for more information.)

David Schatten, of Universal Security Systems, Inc., Hicksville, NY, designed Englewood's electronic access control system. Security officers are kept informed via a clear, graphical monitor display that pinpoints every alarm point and access area by map. The software, which runs on the OS/2 operating system, was chosen for its multi-tasking ability and fast throughput.

To best manage detection devices, the electronic access control system shares alarm point monitoring with a CCTV system and integrates access control and video badging. At this time, over 6,000 badges have been made, with more being added by the day.

The system is based on a very complex network of communications throughout the multi-building campus. It integrates standard hardwired connections, leased lines and wireless spread spectrum reader technology, along with standard readers. The resulting system is seamless, low in maintenance, high in responsiveness and very reliable.

Consolidated Edison of NY

Consolidated Edison of New York
Power Generating Facilities

With over 15,000 cards in the system, not counting those used by visitors and temporary contractors, Consolidated Edison needed a state-of-the-art *accountability system* to augment its emergency procedures in addition to a new electronic access control system.

When emergencies occur in a power plant, people need to know exactly who is where. David Schatten of Universal Security Systems, Inc., Hicksville, NY, together with Con Edison's engineers, developed a unique computerized accountability system to quickly find who is in the facility when there's an emergency.

The key to this system is *speed*. To facilitate this, David specified custom-designed software that groups card users by managerial divisions, with the whereabouts of people in each division being the responsibility of a single person.

During an emergency, managers receive a report of everyone under his or her care who previously checked into the plant. Simultaneously, everyone quickly goes to designated safe havens outside the plant. There, the managers take a roll-call based on their divisional reports, the result of which accounts for who is or is not safe.

In the future, roll-call readers will be added to further speed up the process. Under these conditions, people will swipe their cards through dedicated roll-call readers. Once done, the computer subtracts their names from the divisional reports. Names remaining on the reports, then, indicate who must be found. The goal is to produce blank reports, indicating that everyone is safe.

Getting the system to work properly requires that people follow check-in and checkout EAC rules under normal conditions. These rules are fairly easy to impose at pedestrian gates, where people enter and leave one-at-a-time, but vehicle access poses a problem. This is because a single vehicle entering the facility can often hold multiple passengers.

To guard against procedural mistakes by people in vehicles, security officers are stationed by specific vehicle gates. Every officer carries a wireless laser gun bar code reader and a wireless gate release button. When a vehicle stops at the gate, its occupants must submit their cards one at a time to the officer.

The check-in process audibly and visually confirms access requests. When the laser gun reads a valid card, a pleasant bell sounds and a green light flashes, indicating the card is OK. If the card is invalid, a horn blasts, a red strobe light flashes and the person presenting the card is asked to step out for individual processing so as not to hold up traffic.

Once the officer is satisfied that everything is in order, he or she presses the gate release button, opening the gate. This same procedure is used when vehicles exit the plant.

With all procedures in place and check-in and -out rules are obeyed, everyone is registered in the system. A further benefit is that during an emergency, the system provides speed and accuracy in performing a head count. The resulting accountability gives Con Edison greater security, safety and *superior responsiveness* when the plant is on alert.

Chapter Review - Case Histories

Today's security professional must understand access issues, manage electronic devices and be a computer whiz.

While there are no review questions for this chapter, we encourage you to ponder its stories and to read magazines that also feature case histories. The more information you can soak up from other people's experiences, the better you'll be able to perform your job.

Appendix

Organizations

The following organizations offer professional development in the areas of security and/or building management:

American Society of Industrial Security - ASIS
1655 N. Fort Meyer Drive, Suite 1200
Arlington, VA 22209
703 / 522-5800

Association for Facilities Engineers - AFE
(Formerly the *American Institute of Plant Engineers*)
8180 Corporate Park Drive, Suite 305
Cincinnati, OH 45242
513 / 489-2473

Associated Locksmiths of America - ALOA
3003 Live Oak Street
Dallas, TX 75204
214 / 827-1701

Association of Certified Fraud Examiners - ACFE
716 West Avenue
Austin, TX 78701
512 / 478-9070

Building Owners and Managers Association - BOMA
1201 New York Avenue, NW, Suite 300
Washington, DC 20005
202 / 408-2662

Business Espionage Controls and Countermeasures - BECCA
PO Box 260
Fort Washington, MD 20749
301 / 292-6430

International Association for Healthcare Security & Safety - IAHSS
PO Box 637
Lombard, IL 60148
708/953-0990

International Facility Management Association - IFMA
1 East Greenway Plaza
Houston, TX 77046
713 / 623-4362

International Society of Facility Executives - ISFE
336 Main Street
Cambridge, MA 02142
617 / 253-7252

National Burglar & Fire Alarm Association, Inc. - NBFAA
7101 Wisconsin Avenue, Suite 901
Bethesda, MD 20814
301-907-3202

Information on EAC Equipment Manufacturers

Security Industry Association - SIA
1401 I Street, Suite 1000, Washington, DC 20005
202 / 296-9410

Recommended Books

Source of Security-Related Books and Publications:
Butterworth-Heinemann Publishing & Digital Press,
313 Washington Street, Newton, MA 02158
617 / 928-2500

Connections and Components:
SDM Field Guide, Security Distributing and Marketing, 800 / 662-7776. This is an excellent training manual written in magazine format. It provides insight into the installation of EAC devices and communications issues.

Computers and Networks:
Guide to Connectivity, by Frank J. Derfler, Jr., published by *PC Magazine*.

How Computers Work, Ron White, Ziff-Davis Press.

How Networks Work, Frank J. Derfler, Jr., and Less Freed, Ziff-Davis Press.

The Essential Client/Server Survival Guide, by Robert Orfali, Dan Harkey, and Jeri Edwards, published by Wiley Computer Publishing.

Security-Related Magazines

Access Control and Security Systems Integration, 6151 Powers Ferry Road, NW, Atlanta, GA 30339

Building Operating Management, 2100 W. Florist Avenue, Milwaukee, WI 53209

Corporate Security, 2333 H Street, NW, Washington, DC 20037

Facilities Design & Management, 6160 N. Cicero Avenue, Chicago, IL 60646

Keynotes, 3003 Live Oak Street, Dallas, Tx 75204

Locksmith Ledger, 850 Busse Highway, Park Ridge, IL 60068

Military Engineer (The), 607 Prince Street, Alexandria, VA 22314

National Locksmith, 1533 Burgundy Parkway, Streamwood, IL 60107

Parking Technology, 84 Park Avenue, Flemington, NJ 08822

Police, 515 N. Washington Street, Alexandria, VA 22314

Police & Security News, 15 Thatcher Road, Quakertown, PA 18951

Safety Source, PO Box 365, Stevens Point, WI 54481

SDM Field Guide, 1350 E. Touhy Avenue, Des Plaines, IL 60018

Security Advantage, 2200 Susquehanna Trail, York, PA 17404

Security Dealer, 445 Broad Hollow Road, Melville, NY 11747

Security Distributing & Marketing, 1350 E. Touhy Avenue, Des Plaines, IL 60018

Security Magazine, 1350 S. Touhy Avenue, Des Plaines, IL 60018

Security Management, 1655 N. Fort Meyer Drive, Arlington, VA 22209

Security News, PO Box 460, Salamanco, NY 14779

Security Technology & Design, 850 Busse Highway, Park Ridge IL 60068

Security Technician, 850 Busse Highway, Park Ridge, IL 60068

Today's Facility Manager, 121 Monmouth Street, Red Bank, NJ 07701

Index

Index

Name: _____

Company: _____

Street Address: _____

_____

City: _____ State: ___ Zip: ____

Dealer/
Distributor Name: _____

WIN-PAK
Northern Computers' premiere access control
software package for Windows.

PAK-TIME *FOR WINDOWS*
Northern Computers' time and attendance
software package for Windows.

N-1000-3
Northern Computers' access control panel.

N-750 System
Northern Computers' access control, burglar,
and fire alarm system.

Name: _____

Company: _____

Street Address: _____

_____

City: _____ State: ___ Zip: ____

HID

Receive $50 in direct cash rebate
from the factory when you buy four
or more HID readers from any supplier.

Send copy of invoice to:

HID

14311 Chambers Rd.
Tustin, CA 92780

Valid for dealers and end users.
Limit: 1 coupon per customer.

Name: _____

Company: _____

Street Address: _____

_____

City: _____ State: ___ Zip: ____

NBFAA
**National Burglar &
Fire Alarm Association**

WANTED!
Security Book Authors

Butterworth-Heinemann, the foremost publisher of books for security students and professionals, is seeking new authors to enhance and expand its publishing program in the security field.

If you are interested in authoring a book for Butterworth-Heinemann, please submit a proposal including:

- A description of the book's aim and scope
- A table of contents or outline, and introduction
- A description of the market for the book
  (Who will buy it?)
- A list of the competing books and their publishers, if known
- Your résumé

Submit proposals to Laurel DeWolf at the address below. Please call if you have any questions.

Laurel A. DeWolf
Acquisitions Editor
Butterworth-Heinemann
225 Wildwood Avenue
Woburn, MA 01801-2041
Tel 617-928-2643
Fax 617-928-2640